D0871784

The

Mirror

and Man

The
Mirror
and Man

BENJAMIN GOLDBERG

UNIVERSITY PRESS OF VIRGINIA

Charlottesville

In memory of my beloved wife

HELEN

whose encouragement
made this book possible

CONTENTS

ILLUSTRATIONS

Illustrations

ACKNOWLEDGMENTS

I am grateful to Mrs. Joyce Strauss, Associate Professor of Art History, University of Denver, for making her thesis, *A Mirror Tradition in Pre-Columbian Art*, available to me, and to Dr. John Carlson, Director of the Center for Archeoastronomy, University of Maryland, for providing me with invaluable source information about the history of the mirror in pre-Columbian America.

I am much indebted to my son, Steven Goldberg, Associate Professor of Law, Georgetown University Law Center, whose painstaking review and editing of all the drafts of my manuscript helped smooth the way to its final version. I would also like to express my deep appreciation to Walker Cowen, Director of the University Press of Virginia, whose personal involvement and definitive suggestions helped shape the format of this book.

Finally, I owe a debt of gratitude to the Library of Congress whose resources helped me immeasurably.

INTRODUCTION

The purpose of this book is to show the influence of the mirror on society from earliest times to the present and into the future. It presents the biography of the mirror, showing the many roles it has developed from infancy to maturity. For as the mirror evolved from a pool of water to the modern looking glass, it developed into an instrument of strangely diversified attributes: feared or extolled by some, worshipped or exploited by others, an instrument both for self-revelation and revelation of the universe.

The many attributes of the mirror become evident when we see how it has permeated the mythologies, superstitions, and religions of many cultures and how it has penetrated art and literature. For example, the mirror appears as a symbol in such diverse religions as Shinto of Japan, the idol worship of the Aztecs, and Christianity. Painters like Leonardo da Vinci and Jan Vermeer found the mirror invaluable in their art. Shakespeare used the mirror for self-revelation, while Jean Genet, the twentieth-century playwright, wrestled with its mysteries. The mirror also serves as a device of deceit in the hands of the magician and a tool for truth in the hands of the astronomer. Tomorrow it will harness solar energy for space colonies.

It may be useful here to recall the one great attribute of the mirror: how a ray of light striking the surface of the mirror is reflected therefrom. A light beam impinging on a mirror is reflected like a struck billiard ball that ricochets off the sidewall of a billiard table. The successful billiard player in striking the ball knows that the angle the path of the ball makes with the sidewall of the table before the ricochet (angle of incidence) is equal to the angle the path of the ball makes with sidewall after the ricochet (angle of reflection). Similarly, the angle at which an incident ray of light strikes the surface of the mirror is equal to the angle the reflected ray makes with that surface. The few ancients who were aware of this law were able to gain priestly powers by performing mirror magic that confounded and frightened the naive.

Further, as a consequence of this physical law, the concave parabolic mirror can reflect the parallel rays of the sun to a focus to cause burning. A concave spherical mirror does almost as well. This phenomenon was known in many early cultures, and mirrors were used to start fires. Today, on a much grander scale, they are used to focus sunlight to attain exceedingly high temperatures for technological uses.

To appreciate the way the book is structured, the reader should be aware that man has used the mirror in two modes: turned toward him and away from him. In the former mode it touched his psyche and entered into his superstitions, divination practices, religion, art, and literature. In the latter mode the mirror became an impersonal implement of science and technology. On the earth it has become the eye of the reflecting telescope. Planted on the moon by the astronauts, it assists in important astronomical measurements. Orbiting in space, it will help replenish the earth's energy resources.

The book is divided into two parts. Part one emphasizes the inward-looking mirror, that is, man's emotional reactions as he viewed his reflection. The mirror's aesthetic development is chronicled as well. This takes the reader through cultures around the world to the close of the eighteenth century, when the aesthetic development of the mirror was essentially completed and mirrors had become everyday objects.

Part two tells the story of the outward-looking mirror from the time it was nurtured from infancy by the ancient Greeks to its maturity during the nineteenth and twentieth centuries. It has been within these last two centuries that the mirror has made its major contributions to society in science and technology. But despite the ascendancy of the objective mirror during this period, part two closes by illustrating how the subjective mirror still haunts man's psyche.

PART
ONE

THE MIRROR
AND MYSTERY

EARLY MAN viewed the mirror with awe. A pool of water or a piece of polished metal had the power to show him a duplicate of himself that he identified as his spiritual double, or soul. It revealed to him the phantom form of his being that lived on after he died, preserving his ego and giving him a comforting sense of immortality. At the same time it made him uneasily aware of his eventual bodily death when his soul decided to leave his body. This led to myth, superstition, and folklore regarding the mirrored image as well as the mirror itself. The universality of these beliefs reflects the common concerns of early man.

To this day many people cover mirrors or turn them to the wall after a death in the house. They believe that the soul, projected out of the living person in the form of his reflection in the mirror, may be carried off by the ghost of the deceased, which hovers around the house until the burial. This custom is widespread in Europe, as far away as Madagascar and among sects of Mohammedans and Jews. Some even believe that a sick person should cover the mirror in his room or have it removed, for in the time of sickness, when the soul is apt to take flight, there is a special danger of projecting it out of the body by the reflection.

Among such diverse cultures as the Aztecs, the Finno-

Ugric people, the Zulus, and ancient Greeks, we find the common thread of reflection-death superstitions but in different forms. The Aztecs kept thieves and sorcerers away from their homes by leaving a bowl of water with a knife in it near the door. The intruder, alarmed at seeing his image pierced by a knife, would turn and flee. The Zulus would not look into a dark pool because they thought some creature hiding below would swallow their reflection and thereby their souls, causing them to die. A Cheremis girl of the Finno-Ugric culture after grooming herself in front of a mirror could be concerned about the loss of her soul and would say reverentially, "Take not from me my appearance or image."[1] The early Greeks were obsessed with the superstition. Unless specially privileged, an individual was not allowed to gaze at his (or her) reflection in the water, and dreaming of seeing one's reflection in water was considered a sign of death.

A peculiar variation of this mirror superstition existed among the ancient Chinese. To them the mirror was a charm to ward off the attacks of unseen evil spirits that beset their lives. As described by the alchemist Ko Hung (4th century A.D.), Taoist scholars were in the habit of hanging on their backs a bright metallic mirror about nine inches in diameter so that the invisible evil specters would not dare come near them because their reflections would make them visible, a condition they could not tolerate. Faced with such a situation, they would melt away and withdraw without doing harm.[2] A more popular custom was to hang globular mirrors over the beds in one's house to protect the family. Since the spherical mirror "saw" all around, the evil spirits could not avoid being reflected and so would avoid the place. There was a special modification of this superstition that Taoist priests used in trying to revive a dying patient. They would take the patient's coat, hang it on a fresh bamboo branch, and sling a metal mirror over it. They would then carry it out for a walk in order to cause the departed soul to return to it. The idea was that the invisible soul supposed to be hovering in the vicinity would become visible through the mirror reflection and, having nothing to fear from the familiar coat, might be induced to return to its owner.

The basic reflection-death superstition also took on less deadly forms of interpretation. Such mirror superstitions were concerned with the safety and health of an individual, and in its most benign form it offered rituals to aid lovers. (For does not true love come from the soul?) In northern India there was a warning against looking into a mirror that belonged to other people. When you were visiting someone, you did not look into any of his mirrors; for when you left his house you left part of your soul behind you, caught and retained by the mirror that reflected you. This was particularly dangerous since that part of your soul remained under the power of your host, who could manipulate it to his advantage. The Greeks believed that if a mirror were held before a sleeping man during a hailstorm, the storm would cease; this superstition was prevalent into the twentieth century among the women of India, who wore mirrors in their thumb rings for that purpose. As an embellishment of this idea, some covered their blankets with little pieces of shining glass to ensure safety in slumber. In Jewish tradition a mirror was used as a protection against weak eyes. Some scholars would set a mirror in front of them while writing and occasionally stare into it so that their sight might not be dimmed. When it came to love, the Jews of the fifteenth century had mirror rituals (combined with fertility symbols) for inducing the romantic or passionate kind. A romantic youth who wanted to attract the girl of his desire would have to go through a very precise series of steps to assure success. First, an egg had to be secured on a Thursday from a black hen who had never laid an egg before, and after sunset on the same day it had to be buried at a crossroads. After three days it had to be dug up after sunset and sold. With the proceeds a mirror had to be bought and buried in the same spot at the same time of day. The youth had to sleep on that spot for three nights, remove the mirror, and have his desired one look into it. Then she would love him.[3]

Another mirror ritual was far more intricate. The young man had to purchase a hand mirror at whatever price was demanded. He then scraped some of the pitch from the back of the mirror and in the space wrote the name of his beloved three times. At the opportune moment he was required to hold the mirror in front of two dogs engaged in the sexual act so that their images were reflected in it and soon thereafter have the girl glance into it. Then he had to hide it for nine days in a spot she passed frequently, and when that period was over, he was always to carry it with him. The intention was to excite the girl when she was in his company through the magic power of the sexual act, fixed in the mirror that had been associated with her name and person.[4]

The well-known superstition of seven years' bad luck to the one who breaks a mirror is related to the belief that the image is the soul. When you look into a mirror, it captures your soul, and if you break it you break your soul. This angers the indwelling spirit so much, because it has been hurt, that it takes vengeance on the offender by punishing him with seven years of bad luck. Napoleon Bonaparte shared this superstition. "During one of Napoleon's campaigns in Italy he broke the glass over Josephine's portrait. He never rested till the return of the courier he forthwith dispatched to assure himself of her safety, so strong was the impression of her death upon his mind!"[5]

There is a plausible explanation for this superstition. In the early days of glass mirrors, their manufacture was very difficult; hence, they were exorbitantly priced. To break one meant the loss of an item so expensive that an ordinary person might have to save for seven years to buy another.

This superstition troubled early American slaves, and they developed an ingenious countermeasure to overcome its effect. Their remedy was to place the broken pieces of the mirror in a running brook. The trouble would be washed away in seven hours.[6]

Sir James Frazer, the British anthropologist, found that the Greeks believed water spirits could drag a person's reflection under water, leaving him soulless to perish.[7] It is likely that the Narcissus myth, also dealing with the danger of looking at one's reflection, evolved from this superstition. The danger was death caused by the frustration of inordinate love for one's mirror image.

Narcissus looked at his reflection in a pool and mistook it for another person. He perceived his image as a real person of great beauty with whom he fell in love. At first he did not realize that it was his reflection and was devastated when he finally recognized the image as his own. It was then that death became the only possible solution for him. Narcissus died when he recognized the illusion for what it was but could not escape from the love it had aroused. He died when there was no hope left that his passion could be satisfied.

Narcissism in a milder sense was, no doubt, the driving force in the self-preservation of the ego of primitive man. This identifiable ego, manifested by his reflection, was his soul, or his double, identifiable with his body. His reflection, mimicking his every move, represented a duplicate personality that, by living forever, denied the power of death over the human spirit. The mirror-soul idea persists and has been a perpetual theme for storytellers, poets, and playwrights.

The universal mystery of the mirror also involves divination, the art that seeks to foretell the future and discover hidden knowledge. A special kind of diviner—the scryer—stares into a reflecting surface and "sees" images of distant or future events. How scrying evolved and fared throughout history is intriguing.

When members of an early society looked at their reflections and contemplated their souls, there were a few who, when gazing into a pool of water, the highly polished blade of a sword, or the bottom of a shiny cup, saw something more. In staring at the shiny surface, they would soon see their image dissolve to be replaced by other visions. Perhaps, while away from home, the gazer envisaged his family being attacked and destroyed the following day or saw where his long-lost hunting knife was hidden. If he were wise enough, he would not be frightened by the visions but would try to understand them, relate them to his concerns, and evaluate their accuracy. If he were shrewd, he would cunningly not explain his new and unusual knowledge about his visions but would shroud it in mystery. He called his instrument of reflection a magic mirror that he alone could use to tell others about their destinies. He would attain a devoted following and be proclaimed a holy leader with supernatural powers associated with the gods. This is the way, in all probability, the early scryers attained their priestly prominence and came to be revered by prince and pauper alike.

Throughout history there have been several modes of scrying. The most common forms have been catoptromancy, divining by staring into an ordinary mirror, and crystallomancy, gazing into a crystal ball, the technique used by some psychics and seers today. Three other less commonly known forms of scrying have existed through the ages. They are cyclicomancy, hydromancy, and onychomancy. Strange words, perhaps, but they merely denote reflections from different types of shiny surfaces. *Cyclicomancy* is the technique of scrying in cups filled with water, wine, blood, or some other liquid. This method has been one of the most commonly used in the East. *Hydromancy* uses a natural body of water like a lake, stream, or well as an all-seeing mirror. *Onychomancy* uses the oil or soot-covered nails of a virgin youth. The images seen on the surfaces of the nails give the scryer the answers he desires. This form of divination was used in both Eastern and Western cultures.

Cyclicomancy is the earliest form of scrying found in literature. The biblical Cup of Joseph, "in which my lord drinketh, and whereby indeed he divineth," was, no doubt, used for magical purposes.[8] The tradition of cup divination also goes back to the ancient Persians as related in the earliest recorded scrying story. Firdausi's great epic poem, *Shah Namah* (*The Book of Kings*), centers about a scrying cup that mirrors the world and distant persons. The cup belongs to King Kai Khosrau and appears in the love story of Bizhan and Manizha, heroic figures in Persian mythology. The king, beseeched by the people of Irman to rid their ravished country of wild boars, sends Bizhan and Gurgin, an envious companion, to do so. Through Gurgin's intrigue Bizhan falls in love with Manizha, the daughter of Irman's ruler. Though it is forbidden, she takes him to Turan and hides him in her palace. He is discovered and imprisoned in a pit with Manizha as his attendant. In the meantime Gurgin has returned home, where the lame story he tells rouses suspicion. By means of the divining cup Kai Khosrau discovers Bizhan's whereabouts and dispatches the hero Rustan to deliver him. Earlier, the king comforts Giv, Bizhan's afflicted father:

> *Then will I*
> *Call for the cup that mirroreth the world,*
> *And stand before God's presence. In that cup*
> *I shall behold the seven climes of earth,*
> *Both field and fell and all the provinces*
> *Will offer reverence to mine ancestors*
> *My chosen, gracious lords, and thou shalt know*
> *Where thy son is. The cup will show me all.*

Later the poet tells what Kai Khosrau saw in the cup:

> *He saw the seven climes reflected there*
> *And every act and presage of high heaven*
> *Their fashion, cast and scope, made manifest.*
> *From Aries to Pisces he beheld*
> *All mirrored in it—Saturn, Jupiter*
> *Mars, Leo, Sol and Luna, Mercury*
> *And Venus. In that cup the wizard-king*
> *Was wont to see futurity. He scanned*
> *The seven climes for traces of Bizhan,*
> *And, when he reached the Kargasars, beheld him*
> *By God's decree fast fettered in the pit,*
> *And praying in his misery for death,*
> *With one, the daughter of a royal race,*
> *Attending him; the Shah, with smiles that lighted*
> *The dais, turned his face to Giv and said,*
> *"Bizhan is yet alive; be of good cheer."*[9]

Moving from legend to reality: another early form of scrying, crystallomancy, appeared in ancient Gaul among the druids. Using a crystal called plentz, the archdruid divined the future for his followers. A sample of this kind of magic mirror was found on the plain of Stonehenge, a piece of round rock crystal, three inches in diameter and about one inch thick.

The Greeks combined hydromancy with catoptromancy to determine the fate of a sick person. Standing over a suitably sacred spring, the diviner tied a mirror to a fine cord and lowered it until it grazed the surface of the water. After praying to the appropriate goddess and burning incense, the diviner looked into the mirror and saw the sick person either living or dead and made his pronouncement. The Greeks have carried this scrying tradition into the twentieth century. On the island of Andros girls still hold a mirror over a well if they wish to see reflected in it, from the water of the well, the appearance of their future husbands.

All pre-Christian cultures have practiced scrying in one form or another. Its techniques were absorbed by the growing Christian community on the European continent. Mirror folklore spread and mirror divination became popular. Wondrous qualities were attributed to certain magical mirrors owned by the legendary figures Prester John and Merlin. The former, a mighty Christian potentate of the Indies, apparently owed his wealth and power to a special mirror in his possession whose powers he described in a letter to the Byzantine emperor Manual (1162). The mirror, he wrote, revealed all plots and machinations made against him and all good deeds done for him. To prevent it from being broken or stolen, he mounted this miraculous mirror on an elaborate structure of pillars that could be reached only by ascending 125 steps; each step of the way was guarded day and night by twelve thousand armed men.[10]

Merlin, magician to the court of King Arthur, fabricated a similar mirror. It would show to those permitted to look into it anything that pertained to them, anything that a friend or foe was doing. Merlin gave this mirror to King Arthur whose daughter, Britomart, saw in it her lover, Sir Artegal. Spenser tells this story in *The Faerie Queene*. He describes the power of this magic mirror:

The Mirror and Mystery

Whatever thing was in the world contaynd,
Betwixt the lowest earth and hevens hight,
So that it to the looker appertayned
Whatever foe had wrought or friend had fayned.[11]

Only a very few could have their own divining mirrors, so much of the scrying performed for others in medieval days was done by a group of men called specularii. They traveled throughout Europe providing divination services with the aid of a mirror (as their name implied) for anyone who wanted it. They developed a huge following and are mentioned as early as A.D. 450 in the councils of a synod convened by St. Patrick, which shows their influence had spread as far as Ireland.[12]

As the strength of the church increased, the dominance and credibility of the specularii were challenged because they were in conflict with established religious doctrines, and they represented a threat to the authority of the established order. John of Salisbury (c. 1115–1180), a distinguished churchman, was very much disturbed by scryers, who divined in "objects which are polished and shining, like a kettle of good brass, glasses, cups, and different kinds of mirrors."[13]

John's view was based on his childhood experiences. When he was a boy, he was placed in the charge of a priest to learn the psalter. The priest, who also was a magician, induced John and another boy to try out as scryers. This was in accordance with tradition. From earliest times the use of young boys and girls under the age of twelve as scryers was common practice based on the discovery of many magicians that young children, unpolluted by worldly matters, are very susceptible to seeing images of various sorts while staring into a shining surface. They were, therefore, used by the practitioners of the art who themselves often saw no visions in the mirror, crystal, or other reflecting surfaces but who acted as interpreters of what the child saw. In keeping with this custom, the priest tested the boys for their scrying potential by having them stare at their fingernails, which were coated with holy oil, and then at a polished metal bowl. The other boy saw cloudy images, but John saw nothing except his face. John, failing the scrying test, was thereafter ignored by his tutor while the other lad joined the priest as his apprentice. John was horrified at the priest's unorthodox activities, and his horror increased when he grew up by what he observed of the fate of many of these specularii, a loss of their sight from this unholy activity. In later years John listened to the specularii defend their art on the ground that "they make no offerings [to demons], harm nobody, often do good service by revealing thefts, cleanse the world of evil witchcraft and seek truth that is useful or necessary."[14] John, however, contended that the scryers' invocations preceding their visions were ungodly, calling up evil spirits that inevitably led to a bad end.

The battle between the scryers and the church continued for several hundred years. The church condemned them as disciples of the devil who saw and heard demons when they looked into the mirror. Very much concerned by these practices, Bishop Baldock in 1311 wrote to an official of the archdeacon of London, directing him to investigate the matter of sorcery and magic in the city and diocese, particularly the methods used to predict the future by invoking spirits in fingernails, mirrors, stones, and rings.[15] The noted theologian Nicole Oresme (1330?–1382), bishop of Lisieux, was similarly appalled by the spread of scryers in his time. He was especially angered by the common practice of having the supposed seer use young boys as scryers and of subjecting them to long periods of intense staring into polished surfaces, which often led them to the blindness noted by John of Salisbury. Oresme also commented on the terrible changes that the scryer's face underwent during his conjurations and invocations. Often he seemed to lose his identity. He appeared to be so disturbed mentally that it was small wonder he experienced so many fantasies.[16] In 1398 the Faculty of Theology in Paris recognized these kinds of complaints and condemned the specularii as being of satanic origin.[17] Yet despite the church's condemnation, the practice of scrying continued to prevail in all parts of Western Europe.

A special form of scrying became quite notorious from the fourteenth to the sixteenth century. It dealt with the location of stolen goods and identification of the thief. This form of scrying reached such heights that it was outlawed, but belief in its practice was so strong that many people, including churchmen, defied the authorities. Consider the recorded case of William Sadyngstone, abbot of the house of Augustinian Canons at Leicester. Some of his money had been stolen. Calling his associates together, he demanded that the thief confess. When nobody did, he turned to sorcery. On 20 and 21 September 1439, after certain incantations, he anointed the thumbnail of a boy named Maurice and commanded him to tell what he saw in it. As a result of the boy's vision, the abbot accused Brother Thomas Asty of the theft. The others of his congregation were outraged by the abbot's evil action and reported him to the bishop. Sadyngstone was tried and found guilty of sorcery.[18]

To further condemn scrying, the church associated it with witchcraft, a crime of the vilest sort and punished in the most fearful ways. One edict pronounced: "If ever you have . . . given reward to anyone to raise the devil in order to discover lost goods, you have sinned. If you have looked in to a sword, or basin, or thumb, or crystal (or cause a child to do so)—all that sort of thing is called witchcraft."[19]

Somewhat less controversial was the astrologers' use of the mirror. Since astrology and mirrors were separately used for predicting the future, the combination appeared to be ideal to many soothsayers. The cabalists, whose art was based in part on numerology and astrology, used seven metal mirrors, each of which bore the name of one of the then-known planets. The mirror of the sun was made of gold, the solar metal. It could be consulted with advantage only on a Sunday. The mirror of the moon was made of silver and could be consulted only on Monday, and so on for every day in the week: iron (Mars) on Tuesday, mercury on Wednesday, tin (Jupiter) on Thursday, copper (Venus) on Friday, and lead (Saturn) on Saturday. Copying from the cabalists, many astrologers made divining mirrors from an alloy of the seven metals known as electrum. Since electrum was made from the metals named after the planets, it was thought to be able to absorb and retain certain influences (A bowl or pitcher made of electrum instantly detected the presence of any poisonous body introduced surreptitiously by beginning to sweat on the outside.). To make an all-powerful magic mirror, the electrum had to be made not only with specified quantities of the particular metals but under favorable astrological conditions, thus: the alloying of the various metals had to take place at the moment of conjunction of specific planets. Since the designated conjunctions of the specified planets in our solar system take place within a thirteen-month cycle, that period of time was required to complete an electrum mirror. With the aid of such a mirror you undoubtedly became all-knowing, for it was said that you could see events of the past and present; see absent friends or enemies; and all doings of men, day or night. You could also see in it everything that has ever been written down, or said, and by whom; and you could see anything, however secret it may have been.

Regardless of its association with the witch, the astrologer, and the devil, divination was so much a part of daily life in the pre-modern world that the church never conquered it and had to accommodate to it. Indeed, the church established special forms of scrying that called up only good spirits. This was made possible by using so-called theurgical mirrors, which presented only divine actions. Such a mirror was used in a religious atmosphere and addressed in suitably holy words. The scene was something like this: A glass vase filled with clear water rested on a table covered with a spotless white cloth. A candle was burning behind and on each side of the vase. A virgin child knelt before this theurgical mirror and called on the archangel Gabriel, the bearer of light from the Divinity. The holy man overseeing this event addressed Gabriel, "Allow the good angel, entrusted with this child's care, to show her by means of reflection in this water what it may please God she may discover of the reply solicited." The child, after staring at the water for several minutes, saw the angel and asked the questions for which answers were sought. The angel replied to them in allegorical images or by writing. The virgin child then saw the places, persons, or descriptions that provided the answers sought.[20]

In another beneficent setting, detection of a thief could be carried out by lighting a blessed candle, bringing it near a mirror, and having a virgin pronounce these words: "White angel, holy angel, by thy holiness and by my virginity show me who has stolen this thing." The image of the thief would materialize in the mirror for the eye of the catoptromancer to behold.[21]

While disciples of the devil and aides of the Almighty were carrying out their own brands of scrying, some philosophers of the day were trying to explain this widespread phenomenon of mirror gazing. As early as the fourteenth century the Arabian historian Ibn-Khaldun wrote: "Sight is the most noble of all our senses, and is therefore preferred by those practicing divination; fixing their gaze on a [mirror], they regard it attentively until they see that which they declare. The persons are mistaken in thinking they behold objects and visions in the mirror; a kind of misty curtain intervenes between their eyes and the bright mirror and on this appears the phantoms of their imagination."[22] A metaphysical explanation was given by Jakob Böhme (1575–1624), a German mystic, who believed that when the right person engaged in mirror gazing, a form of spiritual energy radiated from the eyes of the gazer and collected on the surface of the mirror. There it formed a sensitive film in which the astral scenery reflected itself and occasionally revealed past, present, and future events.[23]

With the spread of rationalism during the sixteenth century, men began to think for themselves in natural philosophy, religion, and politics. Believers in the supernatural, however, still abounded, providing plausible explanations for the phenomena they described or witnessed. In fact, the popularity and credibility of scrying in the sixteenth century actually increased and soon reached its highest development, not as a result of the exhortations of a bemused mystic or sophisticated confidence man but under the auspices of a distinguished man of science and confidant of Queen Elizabeth named John Dee.

The Mirror and Man

Dr. John Dee, a scholar of the Elizabethan Age, was an important contributor to the fields of mathematics, geography, and medicine. He wrote brilliantly on Euclidian geometry, made a precise readjustment of the Gregorian calendar, and made important geographic and hydrographic contributions to the explanations of the New World. He was also considered one of the greatest exponents of the art of scrying who ever lived. His reputation was strengthened by his knowledge of the science of optics and supported by a library crammed with treatises and manuscripts on all matters optical, from spectrums to magic mirrors. He was in great demand and divined for the nobility of Europe. One of his magic mirrors, known as the Shew-Stone, survives and is now in the British Museum, which acquireed it in 1966.

Dee was born in London in 1527. Before the age of twenty he was made Fellow of Trinity College, Cambridge. After visiting abroad for four years, he returned to England in 1551 and enjoyed the patronage of Edward VI, who in 1553 appointed him rector of Upton-upon-Severn. In 1555, after Queen Mary ascended the throne, he was arrested on suspicion of favoring the cause of Elizabeth but was lucky enough to be acquitted by the Star Chamber of an accusation of practicing sorcery against Mary's life. He must have helped Elizabeth, for after her accession she favored him highly, as noted by an item in his diary for March 16, 1575: "Her Majestie willed me to fetch my glass so famous, and to show unto her some properties of it, which I did; her Majestie being taken down from her horse by the Earl of Leicester, did see some of the properties of that glass to her Majestie's great contentment and delight."[24]

After 1578 Dee's interests turned to psychic phenomena. He spent the next few years looking for a suitable scryer to work with him and in 1582 hired Edward Kelly, who established himself in Dee's confidence and soon became his closest associate, acting not only as his scryer but also as a close adviser and traveling campanion. The choice of Kelly as a scryer was contrary to all precepts established by his predecessors. Kelly certainly was not an innocent youth, free of sin and unpolluted by the ways of the world. He seems to have been a well-known scoundrel whose ears had been cut off for some misdeeds in Lancashire. His influence on Dee was enormous, and the good doctor was never quite the same again. A verse in the second part of Samuel Butler's satirical poen *Hudibras* (1664) describes the relationship:

> *H'had read Dee's prefaces before,*
> *The Dev'l and Euclid o're and o're*
> *And, all the intregues 'twixt him and Kelly,*
> *Lescus and th' Emporer would tell yee. . . .*
> *Kelly did all his feats upon*
> *The devil's looking-glass, a stone,*
> *Where playing with him at Bo-peep*
> *He solv'd all problems ne'r so deep.*[25]

Dee's reputation deteriorated as a result of his association with Kelly. Yet to the end he was a man of profound piety. He believed that with Kelly as his scryer he was able to communicate with the angels, enabling him to advise his clients what was truly in their best interests.

Dee was steeped in cabalistic lore and employed its symbolism. When he set up his mirror for scrying, Dee placed it upon a waxen block inscribed with a pentagram and many Hebrew words derived from the cabala. The block rested on a brilliantly colored table that was insulated from the floor by specially inscribed waxen blocks under its legs. The inscriptions contained the five-pointed stars of natural earthly magic and the seven-pointed star of the deeper cabalistic mysteries of the Divine Essence. He believed that his black mirror, or Shew-Stone, was a means of testing the spirits for their relationship to God and claimed that through his Shew-Stone he was able to communicate with the spirits of the dead.

Dee's Shew-Stone can be seen today. The story of how it wandered for nearly two hundred years until it found a resting place in the British Museum begins with an antique collection owned by Horace Walpole. In 1784 the contents of

The Mirror and Man

Walpole's Strawberry Hill estate were cataloged and printed. One of the items was "a speculum of kennel-coal, in a leathern case. It is curious for having been used to deceive the mob by Dr. Dee, the conjuror, in the reign of Queen Elizabeth. It was in the collection of the Mordaunts [who acquired Dee's estate], earls of Peterborough. . . . From the Mordaunts, it passed [through two more owners and then] to Mr. Walpole."[26]

In 1842 the whole Strawberry Hill collection was auctioned. John Hugh Smyth-Pigott bought the magic mirror and seven years later sold it to Lord Londesborough for his famed collection. In 1856 Thomas Wright wrote a historical introduction to a book on the Londesborough Collection entitled *Miscellanea Graphica* in which the Dee mirror was illustrated and discussed. The mirror is described as "a polished oval slab of black stone, of what kind we have not been able to ascertain, evidently of a description which was not then common in Western Europe, and Dr. Dee may have considered it as extremely precious and only to be obtained by some extraordinary means."[27]

Before Dee's mirror was finally acquired by the British Museum in 1966, a remarkably accurate guess as to its composition was made by O. M. Dalton in 1906. Without ever seeing the Shew-Stone, Dalton, an associate of the museum, guessed it to be a piece of polished obsidian.[28] This was verified sixty years later when it was in the museum's possession. Obsidian is a naturally occurring black glass and is in abundant supply in the hills of Mexico. The Aztecs made mirrors as well as knives and arrows from this material. Except for producing a black image, the obsidian Shew-Stone's simple appearance belies its reputation. It is about one-half inch thick, roughly circular, about seven and one-half inches in diameter with a short projecting handle containing a small hole in the center. Its overall length is about nine inches. How did Dee get this mirror and what accounted for its reputation?

After the Spaniards conquered Mexico in the early part of the sixteenth century, they brought back many valuable art objects, including an obsidian mirror treasured by the Aztecs because of its association with their god Tezcatlipoca. This deity always carried such a mirror to see what was going on anywhere and anytime in his domain. He was the omnipotent scryer. The Aztecs firmly believed in this form of divination, and their seers used these mirrors to reveal the future of the tribe and the will of the gods. Dee probably obtained his obsidian mirror on one of his journeys to the Continent soon after it reached Spain from the New World. No doubt he heard of its powers from the Aztec tales surrounding it and was sure he had an extraordinary item.

The Aztec civilization was not the only culture in the New World to practice scrying. Mayan medicine men used the *zaztun*, a quartz crystal or other translucent stone which, having been duly sanctified, was endowed with the power of reflecting the past and the future. The Pawnees of the West of the United States used an unusual form of mirror gazing. When a badger was killed, it was kept by the older people until night and then skinned. The blood was poured into a bowl, and the children had to look at themselves in it by moonlight. If they saw themselves with gray hair, it meant long life; if the picture was dark and indistinct, the child would die of sickness; if no image were seen at all, the child would live and be killed by the enemy.

Scrying extended to all populated areas of the world. Despite the lack of linkage between peoples, their concerns, fears, and curiosity appear to have been similar. Divination with shining surfaces offered a look into the unknown. In Australia crystal was a valuable part of the medicine man's equipment. He used it for rainmaking and predicting the results of distant expeditions. In the region formerly known as the French Congo, the natives used a form of scrying in an initiation ceremony. At the back of the ceremonial hut under a roughly hewn statue of a deity were placed the bones of a man long since dead. A mirror was placed in a manner to reflect the bones. The candidate was required to look at the reflection, and until he could see in it and accurately describe the features of the dead man (whom he could not possibly have known), he was not allowed to proceed with the remaining tests awaiting him before initiation.

In Tahiti priests performed a unique form of divination to catch a thief. When a villager was robbed and wished to have the thief identified, he would call upon the priest for help. The priest, arriving at the victim's house, directed that a hole be dug in the floor and filled with water. Then taking a young plantain in his hand, he stood over the hole and offered prayers to the god whom he invoked, who then was supposed to conduct the spirit of the thief to the house and place it over the water. The image of the spirit, presumably of the thief, was then reflected in the water and the thief himself could be sought out and punished.

In India there existed a mode of divination called "viewing of unjun [lampblack]." The unjun was applied to the palm of the hand of a child, and he was directed to stare intently at it to ascertain where stolen goods were hidden or where treasure was buried.

With the establishment of hypnotism as a scientifically acceptable phenomenon by the work of James Braid in 1841 and the later effort of H. Bernheim in 1884, mirror or crystal gazing was considered a form of hypnotism. One investigator noted that the gaze is fixed on a mirror or crystal with the expectations of seeing visions in it; and it is only after gazing until the state of hypnosis (in which hallucinations can be produced) is reached, that visions occur.

Significant experiments in crystal gazing were carried out by an Englishwoman, A. Goodrich-Freer, who presented her findings in a paper before the Society of Psychical Research in 1889. Her purpose was to determine by examination and experiment whatever elements of truth the traditions of mirror gazing may contain and possibly apply them to the understanding of some of the questions of the subconscious workings of the mind that still remained to be learned.

The results she obtained enabled her to classify three kinds of vision: (1) afterimages, or suppressed memories brought back from subconscious strata to which they had sunk; (2) externalizations of ideas, or images consciously or unconsciously in the mind of the gazer; (3) visions, possibly telepathic or clairvoyant, implying acquirement of knowledge by supernormal means.

The author could not offer any explanation for clairvoyance or telepathy and hoped that others would attempt a wider and more systematic investigation for themselves.[29]

Goodrich-Freer's results were supported by the research of Max Dessoir, a professor at the University of Berlin and a student of the conscious and subconscious states of the human mind. His experiments with mirror vision showed him how the earliest memories of an individual can be awakened by the process of mirror gazing. He stressed that when reflections are seen in magical mirrors, crystals, or fluids, the most important factor was the person who saw and not the instrument of seeing. Dessoir believed that this phenomenon did not occur in all persons but seemed to be a faculty of the creative individual.[30]

Perhaps the best definition of scrying is provided by Theodore Besterman: "Scrying is a method of bringing into the consciousness of the scryer by means of a speculum through one or more of his senses the content of his subconsciousness, of rendering him more susceptible to the reception of telepathically transmitted concepts, and bringing into operation a latent and unknown faculty of perception".[31]

There is another explanation that should not be overlooked. Scrying is the technique used by "cunning men," as described by Keith Thomas in his *Religion and the Decline of Magic*.[32] Such persons in the guise of wizard or seer depend on the desires, beliefs, or hopes of their clients to shape their prognostications. If the client finds the seer's remarks plausible, then he is satisfied. This is well understood by the seer, and, for example, when he forecasts the future for his customer, he tries to determine his hopes and aspirations; or when he is trying to locate a thief for a client, he looks for suspicions already present in the client's mind and then acts upon them. Anthropological studies show how the African diviner, despite his imposing costume and strange evocations, usually acts as a vehicle for the expression of suspicions that have already been formulated by his client.[33] Normally he leaves it to the client to name the suspect. The verdicts of the witch doctors must be in line with general expectations. In a similar technique used in England, the diviner would ask the client for a list of suspects, and carry out a series of tests designed to isolate the guilty one, carefully watching for his customer's reaction as each name was pronounced.

In another part of the world, members of all the sects of Tibetan Buddhism are aware of a religious form of scrying done in connection with their legendary king Ge Sar. It is an elaborate ceremony, and certain symbolic preparations are made before starting the act of divination. A painted scroll with a picture of King Ge Sar is hung on the wall and a table is placed below it. Three cups are set, filled with Tibetan beer, milk, and tea, and on the right side of the table a vessel is filled with grain, into which a "divination arrow" has been stuck with its point downward. On the left side of the table another vessel is placed similarly filled and upon which there is a silver mirror. The mirror is then covered with five pieces of silk of different colors. Finally, one, three, or five lighted oil lamps are placed in front of the three cups.

The priest who directs the ceremony is one of great spiritual powers. He takes his seat in front of the table and, after burning incense, honors King Ge Sar with a prayer and asks for his help. Then a boy is led into the ceremonial room and is seated on a white cushion in front of the mirror. He is about eight years old and comes from an upper-class family. Under no circumstance can he be the son of a butcher or a blacksmith. The priest now removes the five covers of silk and the boy is asked to gaze into the mirror. If the scrying is to be successful, the boy will soon claim he sees visions in the mirror. He then describes these visions to the priest, who deduces from the youngster's account the answers to the questions which Ge Sar should clarify.[34]

Scrying on a grand scale plays an important role in the selection of Tibet's spiritual and temporal leader, the Dalai Lama. The divination is accomplished by a holy man who interprets certain images that appear to him on the reflecting surface of one of the many sacred lakes of Tibet. It has long been the belief of Tibetans that visions of the future can be seen on the surface of such lakes. The fourteenth Dalai Lama (born in 1935) relates how he was found (when he was two years old) by this traditional method.

According to this tradition, each succeeding Dalai Lama is a reincarnation of his predecessor. Therefore, upon the death of the thirteenth Dalai Lama two things took place: a high lama was appointed regent to rule the country until the new reincarnation could be located and nurtured to maturity, and holy men were consulted to determine where the reincarnation had appeared.

After determining that the new Dalai Lama should be sought in the eastern part of the country, the regent went to the sacred lake of Lhamoi Latso, about ninety miles southeast of Lhasa. It is the most prominent one in the country. There, after meditating for several days, he envisioned on the waters of the lake a monastery with roofs of jade green and gold and a house with turquoise tiles. This vision was transmitted to the high lamas, who searched throughout Tibet for these structures. They were found in the east in the region of Dokham. First, the holy men saw the green and golden roofs of the Kumbum monastery and soon thereafter they saw a house with turquoise tiles in the village of Takstar. In that humble home the holy men found a two-year-old boy who they determined was the reincarnation they sought and who eventually became the fourteenth Dalai Lama.[35]

Many today still believe strongly in scrying, not as a supernatural event, but as an unexplained form of energy manifested in extrasensory perception or telepathy. Study continues on these mental attributes. Their worth remains to be evaluated.

NOTES AND SOURCES

1. Uno Holmberg, *The Mythology of All Races*, 13 vols. (Boston: Marshall Jones, 1927), 4:12.

2. Friedrich Hirth, "Chinese Metallic Mirrors," in *Boas Anniversary Volume* (New York: G. E. Stechert, 1906), p. 229.

3. Joshua Trachtenberg, *Jewish Magic and Superstition* (New York: Behrmans Jewish Book House, 1939), p. 43.

4. Ibid., p. 128.

5. Thomas S. Knowlson, *The Origins of Popular Superstitions and Customs* (New York: J. Pott, 1910), p. 163.

6. Newbill N. Puckett, *Folk Beliefs of the Southern Negro* (Chapel Hill: Univ. of North Carolina Press, 1926), p. 442.

7. James G. Frazer, *The Golden Bough*, 12 vols. (London: Macmillan, 1927), 3:94.

8. Genesis 44:5.

9. Firdausi, *Shahnama*, 7 vols., tr. A. G. Warner and E. Warner (London: Kegan Paul, Trench, Truber, 1908), 3:317–18.

10. Berthold Laufer, "The Prehistory of Television," *Scientific Monthly* 27 (1928):459.

11. Edmund Spenser, *The Faerie Queene*. 3.2.18.

12. A. Goodrich-Freer, "Recent Experiments in Crystal-Vision," *Proceedings of the Society for Physical Research* 5 (1889):492.

13. Theodore Besterman, *Crystal-Gazing: A Study in the History, Distribution and Practice of Scrying* (London: W. Rider, 1924), p. 47.

14. George L. Kittredge, Witchcraft in Old and New England (Cambridge: Harvard Univ. Press, 1929), p. 185.

15. Ibid., p. 187.

16. Lynn Thorndike, *History of Magic and Experimental Science*, 8 vols. (New York: Macmillan, 1923–58), 3:430.

17. Goodrich-Freer, "Recent Experiments," p. 493.

18. Kittredge, *Witchcraft*, p. 187.

19. Ibid., p. 51.

20. Thomas Welton, *Mental Magic* (London: G. Ridway, 1884), p. 93.

21. Emile A. Grillot de Givry, *Witchcraft, Magic and Alchemy*, tr. J. C. Locke (New York: Dover, 1971), p. 307.

22. Henry C. Bolton, "A Modern Oracle and Its Protypes," *Journal of American Folklore* 6 (1893):37.

23. Besterman, *Crystal-Gazing*, p. 51.

24. Goodrich-Freer, "Recent Experiments," p. 495, n. 1.

25. Samuel Butler, *Hudibras* (1664; rpt. ed. Yorkshire, England: Scolar Press, 1970), pt. 2, canto 3, p. 148, lines 9–12; p. 176, lines 13–14; p. 177 lines 1–2.

26. Hugh Tait, "The Devil's Looking Glass," in *Horace Walpole: Writer, Politician and Connoisseur*, ed. Warren H. Smith (New Haven: Yale Univ. Press, 1967), p. 200.

27. Thomas Wright, *Miscellanea Graphica* (London: Londesborough, Albert, 1857), p. 82.

28. Tait, "Devil's Looking Glass," p. 204.

29. Goodrich-Freer, "Recent Experiments," pp. 504–21.

30. Max Dessoir, "The Magic Mirror," *The Monist* 1 (1980):100–109.

31. Besterman, *Crystal-Gazing*, p. 160.

32. Keith Thomas, *Religion and the Decline of Magic* (London: Weidenfeld & Nicolson, 1971), p. 549,.

33. Ibid., p. 549 n. 2.

34. René de Nebesky-Wojkowitz, *Oracles and Demons of Tibet* (The Hague: Mouton, 1956), pp. 462–63.

35. Dalai Lama, *My Land and My People* (New York: McGraw-Hill, 1962), pp. 21–22.

Other sources for this chapter include: Rudolph Brasche, *How Did It Begin?* (Croydon, Australia: Longmans, 1965); Daniel G. Brinton, "The Folk-Lore of Yucatan," *Folk Lore Journal* 1 (1883):244; Ernest A. W. Budge, *Amulets and Talismans* (New York: Collier Books, 1930); C. A. Burland, *The Magical Arts* (London: Barker, 1966); William Crooke, *The Popular Religion and Folk-lore of Northern India* (London: A. Constable, 1896); Bernard Fielding, "The Occult Lore of the Mirror," *Occult Review* 29 (1919):144–52; William Fielding, *Strange Superstitions and Magical Practices* (Philadelphia: Blackiston, 1945); N. H. Franklin, *Diary of John Dee* (Carversville, Pa.: Hillside Press, 1963); Peter J. French, *John Dee* (London: Routledge and Kegan Paul, 1972); Douglas A. Hall, *Magic and Superstition* (Feltham, England: Hamlyn, 1968); Owen S. Rachleff, *The Occult Conceit* (Chicago: Cowles, 1971); Northcote W. Thomas, *Crystal Gazing* (New York: Dodge, 1905); William S. Walsh, *Handybook of Literary Curiosities* (Philadelphia: Lippincott, 1893).

EGYPT
AND INDIA

THE BIRTH of the man-made mirror is shrouded in the mists of prehistory. The earliest mirrors we know were found in the region of El-Badari, a civilization of predynastic Egypt. Made before 4500 B.C., they were found in the early 1920s by Guy Brunton, the prominent British archaeologist. Digging in the ancient grave sites of El-Badari located near the Nile about 250 miles south of Cairo, Professor Brunton discovered a slab of selenite, a pearly variety of gypsum, with traces of wood around it suggesting a mirror in a frame. Continuing his search, he made a more substantial discovery: a rather large disk of slate, about eleven inches in diameter, one-half inch thick, with a hole near the upper edge for suspension and a personal mark incised at the lower edge. When its surface was wetted by licking or other means, the slate made an efficient reflecting surface. Further efforts by the British archaeological party in Egypt led to the discovery of the earliest metal mirror extant. It was found in a grave site identified as of the First Dynasty (2920–2770 B.C.). It is made of copper, shaped somewhat like an upside-down pear, with a stem for a handle. This mirror, which is about eight inches in width, is in the Sydney Museum, Australia.

The true origin of this mirror has been a puzzle to archaeologists because its singular shape does not conform to

Egypt and India

the traditional Egyptian metal mirror. The Egyptians, who were sun worshippers, invariably chose the solar disk as the design for their mirrors, a slightly flattened circle, representing the sun as it often appears on the horizon due to the distorting effects of the earth's atmosphere. With the exception of this earliest find, metal mirrors did not appear again until the Fourth Dynasty (2575–2415 B.C.), when they were of the solar disk shape. This form prevailed for many centuries. In the Eighteenth Dynasty (1550–1307 B.C.) the pear-shaped mirror occasionally appeared in Egypt, and therein lies the clue to the first metal mirror mystery. During this era there was a vast influx of Asiatic products because of Egyptian conquests, and the pear-shaped mirror was one of the imports. This is verified by engravings of this kind of mirror on early Hittite monuments, and it seems to have been a familiar form in lands beyond Egyptian control.[1] It seems likely, therefore, that the earliest metal mirror found in Egypt arrived via an ancient Asiatic trade route before the Egyptians initiated their own mirror fabrication during the Fourth Dynasty. Despite its likely origin elsewhere, however, the Egyptian mirror was the first to leave a historical trail.

The Egyptians regarded the polished disks of their mirrors as a symbol of the sun, for the mirrors were able to pick up its rays and give off light as did the great life-giver of the universe. This connection with their sun-god, Ra, made the mirror a religious symbol, and it was used during festivals and on ceremonial occasions. Mirrors were placed in the tombs of the dead, often before the face or on the breast of the corpse, to assure the presence of Ra and to provide for the retention of the soul in the resurrection. At Thebes in a tomb of the Twelfth Dynasty (c. 1950 B.C.) there is a mummy on whose breast a mirror rests. The name of its owner is inscribed on the mirror handle of ebony and gold: "the Magnate of the Tens of Upper Egypt, Rany-sonbe, repeating life."[2]

Until about 2100 B.C. mirrors were usually made of copper; thereafter, they were made primarily of bronze. Because

The Mirror and Man

bronze could be easily cast, more delicate craftsmanship followed and graceful handles of intricate designs were fabricated. When the handle is an intrinsic part of the mirror casting, it is more properly called a tang. When the tang is sheathed in gold, ivory, wood, or other material, it becomes a handle. In addition to making mirrors of copper and bronze, the Egyptians foreshadowed the future by making some small mirrors of black glass framed in wood. They reflected very badly, however. Two such mirrors are in the Turin museum. The technique was acquired during the Eighteenth Dynasty from Syria, where glass was accidentally discovered.

The handles of the early Egyptian mirrors are unique. These handles are the earliest examples of symbolism incorporated in mirror design. The most usual form was that of a papyrus stalk, which represented youthful vigor. Sometimes the upper end of the handle was embellished with the head of the cow-eared Hathor, goddess of love and fertility. Occasionally, a figure of the god Bes served as a mirror handle, associating the mirror with art, dance, and music over which Bes presided. With the advent of the Eighteenth Dynasty secular handles appeared in the form of nude serving maids, who held the disks for the use of their owners (fig. 1).

The mirror handle represented youth, love, and grace—an elixir for all. It is not surprising, then, that the mirror was an important beauty accessory for nearly every fashionable Egyptian, man or woman, and was engagingly called a "see-face."[3] It was used to adjust one's coiffure or costume from earliest times, but its most important use was to make possible the skillful application of cosmetics. Archaeological evidence shows that cosmetics were used by men and women of the predynastic Badarian civilization, a strong implication of mirror use.

Cosmetics served a twofold purpose: to beautify and to pprevent dry skin under the hot Egyptian sun. Eyepaint, for example, in addition to giving an illusion of size and luster to the eyes, helped soften the glare of the intense sunlight. Because of the high development of the mirror with its good

Eighteenth Dynasty copper mirror with handle in the form of a nude female supporting a papyrus umbel. (Courtesy of the Oriental Institute, University of Chicago)

reflective properties, the early Egyptians were able to exploit the art of makeup to its highest degree. To appreciate truly the importance of the mirror in this intricate operation, consider what the fashionable woman of the day had to contend with: after applying a yellow liquid to her face, neck, and arms to lighten the skin and give it the right glow, she spat into an elaborate cosmetic spoon and added some green powder from a small pot, mixing the contents to form a green liquid. While the yellow covering was still moist, she dipped her fingers into the green fluid and stroked it across her eyelids, carefully blending the green edges into the yellow base. After the wet surfaces were dry, she moistened a smooth, slender rod in her mouth and dipped it into a small vessel containing a black powder. With the blackened rod she then drew a heavy line across each of her eyelids, just above the lashes, extending it somewhat beyond the outer corner of each eye. Then she painted on long and heavy eyebrows slightly above where the natural ones had been shaved off. When she completed her eyes, the lady applied rouge, a mixture of brownish red pigment and scented ointment, to her cheekbones, blending it carefully into the yellow base. She completed her makeup with rouge to her lips.[4]

The fine art of makeup depended on the presence of the mirror, and its obvious relationship is shown in a number of tomb drawings. Not too much different from today, Egyptian women of long ago carried their mirrors and makeup boxes with them to gatherings and festivities and placed them under their chairs for emergency use.

The Egyptians had a rudimentary sense of the optical properties of mirrors. An Egyptian papyrus in the British Museum tells of a séance given before King Khufu in which a magician replaced a severed head (an optical illusion).[5] In a very practical application, mirrors were used to reflect sunlight into the depths of the pyramids to allow workmen to see what they were doing in the otherwise darkened tombs.

Mirror art flourished during the Eighteenth Dynasty; thereafter, Egyptians continued to make beautiful mirrors until the Persians invaded Egypt in the fifth century B.C.

During the thousand-year span, their art flowed, via the trade routes, to other countries in the Mediterranean basin, but only the Etruscans and the Greeks established significant mirror traditions of their own. The Egyptian mirror as an art form disappeared with the Roman conquest of Egypt in 27 B.C.

The historic trail of the metal mirror originated in Egypt, but evidence suggests that the man-made mirror could have been born in Asia, possibly in India. In 1946 twelve copper and bronze mirrors were found at Mehi in western India.[6] They were from the Kulli culture, which existed before 3000 B.C. From graves that were rich in copper and bronze objects, they are good examples of the metal craftsmanship of that civilization. The most extraordinary find is a copper mirror about five inches in diameter with a four-inch handle, also of copper, representing a stylized female figure in the manner of the clay figurines found at the same site, with bare breasts and arms akimbo, but with the head provided by the reflection of the user of the mirror. This innovation, combined with the detailed metal work at this early date in history, makes the Mehi mirror one of the handsomest ancient toilet accessories. It displays the genius of the native metalsmiths of the period. The use of a human figure as a mirror handle did not appear in Egypt until about 1500 B.C., in the Eighteenth Dynasty.

As in Egypt, cosmetic artifacts were found where mirrors were excavated. In the Indus valley civilization, which flourished abut 2500 B.C., cosmetic pots, hairpins, and combs were found with bronze mirrors; and it appears that for the arrangement of coiffures and the application of cosmetics, mirrors must have been in great demand.

Little material evidence is available regarding the Indian mirror over the next two millennia. But from later objects of art, particularly sculpture, we learn that the mirror became an exquisite instrument that reflected the grace and beauty of the feminine form. In time it became the symbol of lovers. This is best illustrated by statues of Rati viewing herself in a mirror. Rati, the wife of Kama, the god of sexual love, was

The Mirror and Man

the personification of sexual enjoyment. She viewed her feminine attributes with great satisfaction. This motif was reflected many years later, around A.D. 1200, in the sculpture that adorns the Sun Temple of Konarak, with its many figures of love-making maidens.

An extension of this motif is seen in Indian Buddhist iconography. A mirror appears in one of the six hands of the god Vajrananga (fig. 2), the Buddhist god of love. The worshipper seeking favor before this god would have to concentrate on the following imagery: he pierces the bosom of his beloved with the arrow of the lotus bud. She falls to the ground in a swoon whereupon he ties her legs by a chain which is the string of the bow. Then he flings the noose of the lotus stalk around her neck drawing her to his side. Then he frightens his beloved with the sword and subsequently has only to confront her with the mirror by which she is completely subjugated.[7]

This Buddhist use of mirror symbolism is only a hint of how it was used by that faith. When Buddhism gained a foothold in India during the fourth and fifth centuries B.C. and began to spread throughout Asia, its dissemination was aided by the mirror.

The mirror acted as an interpreter of some of the abstruse concepts of Buddhist philosophy. Because the mirror was a reality that everyone knew, it became an easily understood symbol or metaphor to explain the abstractions of Buddhism. Transcendental thoughts such as purity of the mind, the concept of nonself, and the idea that the universe is one and one is the universe were explained by the mirror and its reflection. How the mirror was used by the Buddhist scholar or priest to explain a difficult idea to the devotee depended on the sect of the religion involved. As an example, it has been discovered through metal samples sealed in Chinese images that the heart of their Buddha is a mirror, a symbol of light and truth. An extension of this idea is found in the native Tibetan deity Zhang Blon, who also has a mirror for his heart. When other Tibetan deities are represented by

Vajrananga. (Firma KLM Private Limited. Photo Library of Congress)

their apparel only, mirrors are regularly placed in the position of their hearts.[8]

The mirror also made its appearance in ancient Indian literature known as Jatakas, a collection of stories associated with Buddha in some of his previous births where he accumulated the Buddha qualities of his present life. Gautama Buddha used his experiences to point a moral or embellish a tale. His disciples learned and repeated them. After his death 550 of these Jatakas were gathererd in one collection. This collection is the oldest, most complete, and most important collection of folklore extant. One such Jataka tells of the wise man, Prajnavanta (Buddha in a previous birth), who while strolling in the marketplace, saw a merchant's son arguing violently with a leading courtesan in the midst of a large crowd. The day before, the youth had offered the courtesan 100,000 pieces to sleep with him. She said that she was unable to accommodate him that night because she had already been hired by another, but would entertain him the next evening. She entertained the other man that night.

That same night the merchant's son had a most erotic dream. He diverted, enjoyed, and amused himself with the courtesan the whole night to his heart's content. She, in turn, after diverting, enjoying, and amusing herself with the other man all night long, approached the merchant's son in the morning saying, "Here I am come to entertain you, sir." The merchant's son replied, "I diverted, enjoyed, and amused myself with you in my dream to my heart's content, the whole night long. Go away, I don't want you." She said, "If, sir, you diverted, enjoyed, and amused yourself with me in your dream to your heart's content the whole night long, you should give me the 100,000 pieces." The merchant's son said, "Seeing that you lay with another man the whole night, why should I give you 100,000 pieces?" The courtesan insisted otherwise, and so the dispute arose between them. A great crowd had gathered but no one was able to settle the argument.

The townspeople, seeing Prajnavanta, appealed to him saying, "What seems to you the proper thing? Should the courtesan be given the 100,000 pieces by the merchant's son or not?" He replied, "The fee should be paid by the merchant's son to the leading courtesan in just the same fashion as he consorted with her." They asked how this could be done, whereupon Prajnavanta ordered that a large mirror and 100,000 pieces should be brought. He told the merchant's son, "Take the box containing the 100,000 pieces and set it in front of the mirror." Then he said, "Come, lady, take this reflection of the box containing the 100,000 pieces which is in the mirror. That is your fee." The crowd roared their approval of the solution found by Prajnavanta. The reflection in a mirror, being as unreal as a dream, can serve to pay in full for a dream.[9]

Reflection used as payment appeared in an unusual Indian ritual over two thousand years later. Travelers visiting India in the eighteenth century reported that women waved mirrors before the image of the goddess Kali to appease that strange and sinister deity. The waving was interpreted as a means of offering the reflection of a person to satisfy the appetite of the goddess rather than a true human sacrifice—a solution that Prajnavanta would have been proud of.

The tradition of the mirror in India does not lie so much in the artifact itself as in its metaphoric application. Its use in Buddhist philosophy helped spread the faith to many parts of the Orient. Of prime interest was the infiltration of this philosophy into the substantial mirror traditions of China and Japan. This is the most important Indian contribution to the history of the mirror.

NOTES AND SOURCES

1. Flinders Petrie, *Objects of Daily Use* (London: British School of Archaeology in Egypt, 1927), p. 29.

2. William C. Hayes, *The Scepter of Egypt*, 2 vols. (New York: Harper, 1953), p. 241.

3. Ibid.

4. Richard Corson, *Fashions in Makeup* (New York: Universe Books, 1972), p. 7.

5. Albert A. Hopkins, ed., *Magic: Stage Illusions and Scientific Diversions* (New York: Munn, 1897), p. 1.

6. P. Singh, *Burial Practices in Ancient India* (Varanasi, India: Prithivi Prakashan, 1970), p. 36.

7. Benoytosh Bhattacharyya, *The Indian Buddhist Iconography* (Calcutta: Firma K. L. Miskhopadhyay, 1968), p. 115.

8. Alex Wayman, "The Mirror as a Pan-Buddhist Metaphor-Simile," *History of Religions* 13 (1974):266.

9. Alex Wayman, "A Jotting on the Mirror: Those of Ladies," *Mahfil* 8 (1971):209–13.

The Mirror and Man

Other sources for this chapter include: John Baines and Jaromir Malek, *Atlas of Ancient Egypt* (New York: Facts on File, 1980); Guy Brunton, *Qau and Badari I* (London: British School of Archaeology in Egypt, 1927); Sudhanshu Chowhury, *Konarak* (Calcutta: Jiten Bose, 1956); Stuart Piggot, *Prehistoric India to 1000 BC* (London: Cassell, 1950); Elizabeth Riefstahl, *Toilet Articles from Ancient Egypt* (New York: Brooklyn Museum, Brooklyn Institute of Arts and Sciences, 1943); Bruno Schweig, "Mirrors," *Antiquity* 15 (1941):257–68.

CHINA

THE TRADITION of the mirror in China is a magnificent one. It began with the Shang dynasty, about 1200 B.C., and developed over the next twenty-one centuries. During those centuries, the Chinese mirror attained through its artists' creativity and artisans' skill an unparalleled level of beauty. The mirror was used in the conventional manner, but it was also revered as a symbol of the universe and highly valued as a talisman, and it was often given as a gift on special holidays, weddings, and occasions of state. The mirror's significance was evident from the designs on the back. They portrayed a range of concepts, from cosmic philosophy to fidelity in marriage. The artistry of the design was the signature of the high culture of the era. The Chinese were the first to record an understanding of the optical properties of mirrors. Unfortunately, the glorious twenty-one centuries of the Chinese mirror ended about A.D. 900 with the end of the Tang dynasty. After that, very little was added to the art, and so this chapter deals primarily with events up to that time.

Mirrors cast of bronze were among the earliest products of Chinese industry. They were produced in An-Yang, the capital city of the Shang dynasty where the earliest Chinese mirror extant was found. The Shang mirror is cast in the form of a disk. It has a slightly convex reflecting surface and the back is decorated with a design of striated quadrants bordered by a band of scallops. The manufacture of mirrors appears to have taken place as early as 1500 B.C. in the Minusinsk Basin of Siberia, the main center of that civilization. Like the Chinese mirror, these Siberian mirrors were usually disks, slightly convex, with a loop on the back side and without ornament. We do not know whether China was influenced by Siberian imports or whether the Chinese originated these mirrors and exported them to its neighbor. Knowledge about Siberian mirrors is very limited. There is some evidence of manufacture to the third century B.C. and, sporadically until the peak of the Mongol empire, from the thirteenth to the fifteenth century.[1]

The mirror's artistic value became evident in 673 B.C., during the Chou period. The emperor found it suitable to give his queen a gift of a "large girdle with a mirror in it."[2] It was also during the Chou period, about the middle of the fourth century B.C., that the earliest recorded use of mirrors for toilet purposes appeared. In contrast to the customs of other nations, Chinese men wore elaborate headdresses that often needed tidying and straightening. It was this masculine requirement rather than women's needs that contributed to the manufacture of reflecting surfaces of high quality. To appreciate the quality of mirrors made as early as the third century B.C., consider that an encyclopedia written at that time, the *Tien-Chung-ki*, stated that with the polished mirror one could examine "the finest hairs" at the temples and eyebrows.[3] Toilet mirrors were restricted to the very highest society and, of course, the grand ladies of those early days who also used them for coiffure arrangement and makeup. These mirrors were round and usually were cast with a pierced boss, or knob, in the middle of the back through which a cord was passed for holding it in the hand.

Mirrors produced from about the fifth century until about 200 B.C. were of high artistic quality and are generally labeled Huai style. Mirror finds of this era have been few except for a particular type known as Shou-chou, named after the town in the Huai River valley where they were found. Discovered in great quantities, they appear to have all

been made from 241 to 222 B.C. After this period the production of the Huai mirror declined, and with the establishment
of the Han dynasty, 206 B.C.–A.D. 220, these were superseded by a new kind of mirror of superb craftsmanship. The
Han mirrors were of the highest quality and represented a
casting technique that seems unbelievably sophisticated for
such an early date. Highly articulated decorations made up
the back side while the reflecting side was beautifully polished. Much like the modern alloy speculum, their white
bronze composition consisted of copper, tin, and lead.

It is not known how the finished mastery of design and
the casting technique of that period were attained, but, no
doubt, it was to a large extent the result of many centuries of
artistic and technical development within the country. Unfortunately, the specific details of this internal contribution
were destroyed in the wholesale burning of books by the
emperor in 213 B.C. What is known is that in 126 B.C., after
the return of General Chang Ch'ien from his celebrated expedition to western Turkestan and Bactria, opening up western
Asia to the Chinese, hundreds of objects, ideas, and methods
previously unknown in China were quickly absorbed into its
culture, including the decorative art of the mirror.

His efforts also contributed strongly to the unification of
China that took place during the early part of the Han dynasty. During this era the country was truly united and
became a powerful empire, permitting the establishment of a
strong individual culture whose traces remained until the last
dynasty in 1911.

An insight into this culture is provided not only by the
craftsmanship displayed in mirror making but also by the
way the people viewed the mirror. The roundness of the
mirror symbolized the canopy of heaven. A mirror was often
hung from the ceiling of a temple during certain rites and
ceremonies as a heavenly representation. The brilliance of
the mirror reflected the intelligence of heaven and provided
the means by which fire was drawn by the sun and water
from the moon. These associations imbued the mirror with a

spiritual quality, acting as the supernatural agent of the sun, moon, and stars, which were believed to control the destiny of man.

Cosmic ideas were symbolized on the backs of many of the Han mirrors by representing a universe of space and time in miniature. The simplest way of indicating space was by showing the Five Directions: the four cardinal points of the compass symbolizing the Four Quadrants of the Vault of Heaven and the center, which was considered the palace of the ruler, or China. The center was designated by the high, round, pierced boss. South, which was always shown upper-most, was represented by the Scarlet Bird, west by the White Tiger, north by the Black Tortoise, and east by the Green Dragon. The directions were also associated with the four seasons, the four elements, and the four creature types:

Quadrant	Supernatural animal	Season	Element	Creature type
S	scarlet bird	summer	fire	feathered
W	white tiger	autumn	metal	hairy
N	black tortoise	winter	water	shell covered
E	green dragon	spring	wood	scaly

These symbols were also linked with the Yin-Yang philosophy, which teaches that two opposing forces underlie nature. The positive force, Yang, represents light, heat, activity, and spiritual things. The negative force, Yin, symbolizes darkness, cold, passivity, and material things. It is believed that these two forces can exist in perfect balance everywhere in order to achieve order, peace, and harmony in all things throughout the universe. To this end, south is considered the region of pure Yang, represented by the scarlet bird, while north is thought to be pure Yin, in the realm of the black tortoise. East is considered to have more Yang than Yin, and west more Yin than Yang. The center represents China, which marks the axis of balance between the two vital forces.

Another form of mirror decoration symbolized space and time in a more sophisticated manner. Space was shown by symbols composed of combinations of straight and broken lines. These stood for eight directions, the four cardinal points and the intermediate ones, northeast, southeast, northwest, and southwest. Time was conveyed by using the twelve symbols of the far eastern zodiac, which were often twelve animals. These twelve animals symbolized the twelve double hours of the day and the twelve months of the year; sometimes twelve Chinese characters were substituted for them (fig. 3).

Bronze mirror, Han dynasty. (Courtesy of the Freer Gallery of Art, Smithsonian Institution, Washington, D.C.)

The Mirror and Man

The full representation of the symbols is:

Animal symbols	Lunar month	Compass direction	Season	Time of day	Dualism
rat	11	N	winter	midnight	max. Yin
ox	12			2 A.M.	
		NE	start of spring		
tiger	1			4 A.M.	
hare	2	E	spring equinox	6	equality
dragon	3			8	
		SE	start of summer		
serpent	4			10 A.M.	
horse	5	S	summer solstice	noon	max. Yang
sheep	6			2 P.M.	
		SW	start of autumn		
monkey	7			4 P.M.	
cock	8	W	autumn equinox	6	equality
dog	9			8	
		NW	start of winter		
pig	10			10 P.M.	

These "cosmic mirrors," as they have been called, have been named *TLV* mirrors by scholars of the Orient because of certain critical markings which resemble these letters. The *L*'s stand for the four points of the compass and the equinoxes and solstices; the *V*'s represent the four corners of the universe and the beginnings of the four seasons. The meanings of the *T*'s are less understood, but Schuyler Cammann, an eminent student of Chinese mirrors, believes that the *T*'s denote screens inside the gates leading into the city of the emperor.[4]

Celestial influences, represented by decorations on the Han mirror, were often invoked. It was believed that the astrological symbols propagated cosmic forces of benevolence, and through them one might acquire wealth, property, peace, advancement to high state position, many descendants, a long life, and other events of good fortune. These thoughts were often inscribed on mirrors given as gifts to friends, relatives, or officials of state. The inscriptions, which sometimes included the date of manufacture, were often also self-serving. They described the craftsman's skill and efforts and the great fortune awaiting anyone buying his product.

The inscriptions quoted below are from mirrors in the Museum of Far Eastern Antiquities of the Hallwyl Museum in Stockholm and from the Axel Lagrelius collection, which together form the most comprehensive Chinese mirror collection outside of China and Japan.

The mirrors from which the inscriptions are taken are numbered as they appear in Bernhard Karlgren's outstanding work "Early Chinese Mirror Inscriptions": Mirror no. 7: "May you forever have sons and grandsons; may you be a prominent person received in Imperial Audience."[5] Mirror no. 82: "May your every day have delight; may your every month have wealth; may you enjoy being free from [bad] events; may you constantly have your will (what you wish); beauties shall assemble [round you], flutes and lutes shall wait upon you; your business shall be prosperous; all things shall be peaceful; the old ones shall revert to being vigorous people; they shall again live in repose."[6] Mirror No. 139: "From the three auspicious metals I have made the mirror; I have my own method (secret process); it eliminates what is baleful; it is suitable for the market; may it cause people to have long life and not know aging; may you be the father of five sons and four daughters, nine children; may your sons be kings and princes; may your daughters be princesses; may your longevity be like that Tung-sang-fu and Si-wang mu."[7]

The significant change in cultural and artistic values that took place from the birth of the Han period in 206 B.C. to its end in A.D. 220 is seen in changing attitudes toward the mirror. The early Han period was a time of growing prosperity. It had a beneficial influence on art because of the perfection demanded by the wealthy and educated classes. This encouraged designers and craftsmen to do their best and is exemplified by the elaborate mirrors of that period. Later, around A.D. 25, the influences of Confucianism changed values. Extravagance and luxury were regarded as vulgar attributes not to be indulged in by gentlemen. As a result, a great number of new mirror types that lacked the high quality of the earlier Han period were produced. Despite this shortcoming, the designs on the backs were vital, and they expressed many new ideas, ideas that were the product of the lower classes who took the lead and made their impact felt in increased symbolism. They sought a supernatural rather than a purely aesthetic appeal.

This trend reached its height during the last days of the Han period when the country was torn by rebellion and warfare and the people were exposed to ever-increasing hardships and danger. Under these difficult conditions the supernatural powers with which the mirrors were believed to be imbued became even more important. Then the most favored mirrors were made not for an elite but for the great mass of people. Vulgarity in design reigned and low standards of workmanship prevailed because the people wanted a charm to protect or guide them. The demand was great and mirrors were produced accordingly. The backs of these mirrors were covered with an increasing number of divinities, and with the fall of the dynasty the cosmic abstractions that the mirror had before portrayed disappeared. The mirror, traditionally invested with a magical power, assumed a more powerful occult role. With this development, the ancient forms of belief disintegrated. There followed a much-weakened religious discipline that opened the door for the acceptance of Buddhism. This religion then permeated Chinese philosophy.

The Han dynasty collapsed after A.D. 220. After nearly four centuries of civil war, foreign invasion, and general chaos, the empire was reestablished with the Sui dynasty, 586–618. It found its golden age during the Tang dynasty, 618–907.

The bygone artistry of the Chinese mirror was revived during this period. As in the Han period, the early years of the Tang dynasty were characterized by considerable wealth and prosperity. The demand for luxury items was great. This stimulated the production of expensive and beautiful objects and resulted in the formation of artistic centers throughout the country, one being Yang-chou, at the junction of the Yangtze River and the Grand Canal, where exquisitely cast bronze mirrors were manufactured. They exhibited high artistic creativity, comingling native and foreign motifs as a result of extensive foreign connections. The designs of these mirrors ranged from the simple to the elaborate, which completely filled the available space with fine detail, from low reliefs to strong three-dimensional effects, and from a deliberate separation of motifs in distinct concentric zones to a merging of the design from the central area to the rim. The mirror sizes ranged from two and one-half to ten inches in diameter (fig. 4).

Bronze mirror, Tang dynasty. (Courtesy of the Freer Gallery of Art, Smithsonian Institution, Washington, D.C.)

Poems lauding the wonderful qualities of the mirror were sometimes inscribed on the backs of the mirrors. One of these, composed by the poet Chien Wen Ti of Liang during the Sui ear, follows:

> *Fair is the magic mirror*
> *Wondrously perfect in its inspired craftsmanship*
> *Its clarity resembles still waters*
> *Its purity seems to mount skyward*
> *Its brightness bathed the halls of the Chin*
> *Its reflections illumined the palace of Chin*
> *It guards against lewdness and draws together the*
> *   altars of the soul;*
> *It responds to all things without fail*
> *High let its writings be hung in jade-like seal*
> *   characters*
> *Long will endure the carvings of the green bronze.*[8]

The beautiful bronze mirror shown in figure 5 represents one kind of mirror popular during the Tang period. With its four masses of main figures, it sharply delineates a plain background. Foreign influences are evident in the parrots and grapes, which were most likely Persian, and the eight lobes of its border, a popular Tang convention probably suggested by the eight-petaled lotus, a symbol of the Buddhist faith derived from India. The distinctive decoration identifies this as a marriage mirror used to protect the bride and groom. The marriage motif is clear. The central portion shows, on opposite sides of the boss, a mythical male and female lion and parrot mates at the bottom. The lobes of the rim contain opposed pairs of sparrows and butterflies, also mates, together with twin buds suggesting flowers that will bloom to fulfillment together. The crane at the top of the central field is a common symbol of longevity, conveying the wish that the married couple might live to a ripe old age, perhaps even enjoy immortality together. These are all natural symbols of a couple in the balanced harmony of an ideal marriage, one of love and fidelity.

China

Some marriage mirrors had magpies as part of their decoration. These go back to an old legend that tells of a mirror owned by a loving couple which broke when they were obliged to part. Each of them took one of the broken halves as a promise of faith. But afterward, when the wife broke her promise of fidelity, her part of the mirror changed into a magpie that flew to her husband to tell him about it. The magpie symbol remains a threatening reminder of marital infidelity.

The legend aside, the veneration of the mirror as a symbol of fidelity is derived from the introspective reasoning of ancient philosophers. They believed that the mirror reveals to him who looks upon it the mask with which he faces the

Bronze marriage mirror, Tang dynasty. (The University Museum, University of Pennsylvania)

world. This self-study leads to self-knowledge, and self-knowledge to self-control, faithfulness, and steadfastness.

The ancient poets, too, sang their praises of these mirrors. Of poignant pathos is a poem by Li Po:

I

Bright, bright, the gilded magpie mirror,
Absolutely perfect in front of me on the jade
  dressing stand.
Wiped, rubbed, splendid as the winter moon;
Its light and brilliance, how clear and round!
The rose-red face is older than it was yesterday,
The hair is whiter than it was last year.
The white-lead powder is neglected,
It is useless to look into the mirror. I am
  utterly miserable.

II

When my Lord went away, he gave me this precious
  mirror coiled with dragons
That I might gaze at my golden-threaded dress of
  silken gauze.
Again and again I take my red sleeve and polish
  the bright moon,
Because I love to see its splendor lighting up
  everything.
In its center is my reflection, and the golden
  magpie which does not fly away.
I sit at my dressing-stand, and I am like the green
  Fire-Bird who, thinking of its mate, died alone.
My husband is parted from me as an arrow from the
  bow string.
I know the day he left; I do not know the year when
  he will return.
The cruel wind blows—truly the heart of the
  Unworthy One is cut to pieces.
My tears, like white jade chop-sticks, fall in a
  single piece before the water-chestnut mirror.[9]

Among the mirror decorations that appeared during the Tang period was one of lions and grapes that deserves special attention. It appeared for the first time early in the period, but it suddenly disappeared before A.D. 900, less than three hundred years later. The meaning of the lions and grapes and the reason for its disappearance are inextricably intertwined.

These mirrors were of the cosmic sort, so popular during the Han period, and were a survival of the TLV symbolism. The basic change was the replacement of the four mythological creatures, which marked the four directions in space, by lions, with a fifth lion sometimes in the center. According to Schuyler Cammann, the lion symbolism was motivated by a popular musical performance at the Chinese court during the Sui and Tang periods. Known as the "Dance of the Lions of the Five Directions," North, East, South, West, and center, it was accompanied by the "Music of Universal Peace."[10] The first performance of the dance, which was adopted from western Turkestan, was given in 577.[11] Since the lions of the dance were directional symbols, they could, very naturally, have been interchanged with the four creatures on the cosmic mirrors. But why use lions to replace wonderful mythical animals like the scarlet bird or the green dragon? In western Asia lions had long been considered solar animals, representing the power and glory of the sun. Tang literary references show that the Chinese also associated them with light. They were described as having brilliant coats and eyes that flashed like lightning. Furthermore, the lions were connected with the "Music of Universal Peace," and their dance signified the final unity, stability, and peace the empire had achieved after nearly four hundred years of troubled times. Therefore, the idea of a world at peace, and symbolic depictions of this notion, would have been especially welcome.[12]

The profusion of grapes in the pattern also related to the mood of the times. The Tang court imported new kinds of grapes from Central Asia. About the year 640 the Chinese learned from the people of Turkestan the art of fermentation to make wine. Soon grapes became symbols of luxurious eating and drinking, making them fitting representations of the riches and plenty of that era.[13]

Underlying the motivation for the lion-and-grape pattern was a religious movement known as Manichaeism, the "Religion of Light," a faith that attracted many Chinese followers during this period. Manichaeism began in Babylon in the third century A.D. Its doctrine became so popular that it spread from Babylon to the Atlantic Ocean and to the China Sea, gaining many illustrious converts, including the youthful Augustine, who later turned to Christianity and became one of its greatest teachers. Manichaean tenets defined light as good and darkness as evil, and its cosmology was represented by the "Five Bodies of Light." The five lions on the mirror, given their association with light, could represent that concept. According to Manichaean teachings, the only way in which universal peace could be achieved was by the final victory of the forces of light and the subsequent complete recovery of the five bodies, or elements, which had been partially lost to the spirits of darkness. If the lions indeed represented the five elements, as seems likely, then their triumph in a cosmic dance would also have represented the ultimate victory of light in the universe.[14]

Grapes also appeared in Manichaean teachings, not as a source of wine, but as food for the devout. The Manichaeans wrote of the "Fruits of Light," which formed the only suitable food for the pious. Grapes were one of these fruits.[15]

With light as the fundamental force of Manichaeism, it is not surprising that the Chinese followers of that faith would have used mirrors to celebrate the "Religion of Light" since mirrors in general had been considered by the Chinese as associated with light, lightning, and solar brilliance for centuries before the Tang period. The curious coincidence is that the lion-and-grape mirrors and the dance of the lions flourished during the brief period when Manichaeism was recognized in China and died out precisely when it was officially destroyed. In 845 the Taoist emperor at that time ordered a religious repression of all foreign religions. His action was aimed primarily against the growing influence of Buddhism and inadvertently banned the increasingly popular religion of Manichaeism as well. Outlawed, Manichaeans and non-Manichaeans would have avoided the use of the lion and grape symbols. This may explain why these patterns were never revived.

Another Tang mirror, a colored cast of which is in the Victoria and Albert Museum, assumed religious significance. It has the typical decorations of that period, but it is much larger than the usual Tang mirror—being about twenty-six inches in diameter. It was found broken in four pieces, in the Haram, near the Dome of the Rock in Jerusalem, and it is now preserved in that mosque. Muslims have long regarded it with great veneration, calling it the "Shield of Hamza," although considering it to be of Jewish origin. Probably it found its way to Palestine through Persia, perhaps in the course of the conflicts between the Turks and Persians in the middle ages.

The ancient Chinese book *Pokutulu* notes that the artist "Huangti cast in metal, sacred vessels, among which were fifteen mirrors. In making these mirrors he put into them the vital essence of creation, therein following the fundamental principles of the universe, so that these compare in brilliancy to the sun and moon, and communicate the will of the gods, thus defending us from evil spirits, and curing our disease."[16] Except for the last phrase, this thought epitomizes the reverential attitude of the Chinese toward the mirror as it is portrayed in so many of their creative decorations. The phrase, which appears almost as an afterthought to the grand idea expressed, characterizes the everyday fears of a superstitious people. The Chinese, like other people, were deeply concerned with illness and dangers, which they believed were brought on by invisible evil spirits existing everywhere. The mirror, which possessed the "vital essence of creation," was able to ward off these malignancies because it was imbued with imaginative qualities that could detect or prevent illnesses.

Several accounts are on record that describe mirrors

which produce effects similar to X-rays. Apparently they could light up the "five viscera" of the human body. The emperor Shih Huang-ti around 220 B.C. claimed that he had such a mirror, which was called *chau-ku-pau* (the precious mirror that would illuminate the bones of the body).[17] In another account Chang Ch'ang, a famous scholar and official who died in 48 B.C., was reported to have obtained an old metallic mirror, more than a foot in diameter, that could light up his bedroom without his using either lamp or candle. After he got it, he never had a day of sickness.[18]

Old mirrors of metal were used to produce certain medicines. After these mirrors were found useless because of deteriorations or breakage, Chinese doctors put them to good account by grinding them into powder and mixing them with other medicines. The ground-mirror potion was prescribed to smooth irregularities in menstruation, to facilitate childbirth, and to cure heart trouble. If it should happen that vermin of any kind had entered the ears and nose, then these organs were to be rubbed with the potion and the vermin would come forth immediately. The standard medical work of that time, *Pou-ts'au-kang-mu*, has a whole chapter on this.[19]

The power of the mirror to ward off evil influences from the living was extended to the dead. Mirrors were placed in graves, generally face-upward on the deceased. There the mirror would protect the body from the marauding spirit and furnish light in the darkness of the tomb. A mystery is associated with this. When the tomb of King Ai of Wei, who died in 296 B.C., was opened, among its contents were "several hundred iron mirrors" placed there to protect him.[20] Why these mirrors were made of iron is not known. Perhaps this is related to the magnetic properties of iron, which the Chinese were aware of. If so, the mirrors would thus be endowed with additional magical properties. In any event, though a few others have been found, references to or descriptions of the use of iron mirrors have not been found in Chinese literature.

The evil influences in life and death of the early Chinese were presumed to be caused by malignant spirits or specters known as *sie* or *mei*, who were often in the form of grotesque animals. The incomparable virtue of the mirror was to unmask these specters. This was the subject of a pamphlet, *Record of His Antique Mirror*, by Wang Tu of the Sui dynasty. He tells how he received an extraordinary mirror from Heu Shing, a scholar of note, who declared, "Wherever you bear it in your hands, hundreds of *sie* will run away from men."

Wang Tu set out on a journey in the second year of the Ta yeh period, A.D. 606, with his valuable acquisition and soon found an opportunity to test its miraculous power. At an inn the innkeeper complained to him of a slave girl of great beauty, whom a lodger had abandoned because she was strangely ill. Suspecting something odd about her abandonment, Wang Tu thought she might be a *mei*. When he fetched his mirror, she screamed pitiably, "Do not kill me," showing herself immediately in her spectral form. Wang Tu put away the mirror, and the girl prostrated herself before him, confessing that she was a vixen a thousand years old who lived by a large willow in front of the temple of a mountain god. Having committed a crime, and to avoid this god's punishment, she escaped, assumed a human form, and married. After traveling for years with her unsuspecting husband, she arrived in this inn where her illness drove her husband away. Now, discovered and deserving to die, she asked to do so in a state of drunkenness. Wang Tu called for wine and invited the whole neighborhood to witness the event. In a short time the beauty was so intoxicated that she began to dance and sing ever more wildly and, after changing into a vixen, died.

Later that year Wang Tu became governor of the province of Ho-peh. A plague broke out there as a consequence of a death. Here was a good opportunity for his mirror to do some good. Its first feat was the restoration to health of the whole family of an underling of the mandarin by simply pointing the mirror toward them when it was dark, for the specter-dispelling mirror regularly emitted a brilliant light in the dark. The patients felt their burning fevers subside as the light from the mirror, which was damp and cool like moonlight, fell upon them. Subsequently, many other people were cured by it in the same way.

Wang Tu lent his mirror to his younger brother, Tsih, because the latter had to take a journey. Tsih returned after three years and gave an account of his adventures. In one adventure, while crossing the Yangtze, his shining mirror had calmed the waves, which a furious gale swept up; in another he treated three maids possessed of demons. The morning after the mirror had shed its light on them, the demons had left their hosts and were found dead. They were *sies*. Two were old rats with heads like wolves. They had long tails and were bald and toothless. The other was a multicolored lizard with two horns.[21] This tale reveals a number of notions the Chinese had regarding the power of the mirror over evil influences.

The supposed magical qualities of Wang Tu's antique mirror were due in reality to an astonishing property it possessed. When Wang Tu received this mirror from Heu Shing, the scholar also said. "Whenever the sun shines on this mirror, the ink of these inscriptions [on the back] permeates the images which it reflects."[22] In other words, if the direct rays of the sun are allowed to fall upon the front of such a mirror and then reflected on a screen, the design on the back will appear to shine through the body of the mirror and appear on the screen (fig. 6). The Chinese called these mirrors *theon-kouang-kien* (mirrors that let light pass through them) because that was the way the phenomenon appeared to occur.[23] Investigation of this phenomenon during the nineteenth century determined that it resulted from a mirror-polishing procedure used by the ancient Chinese and, later, the Japanese.[24] When the slightly convex surface of the mirror was polished, it took on imperceptible irregularities that corresponded exactly to the inscription or design on the back side of the mirror. As a consequence, the part corresponding to the raised pattern on the back became relatively concave and so concentrated the light there more strongly than on the remainder of the slightly convex surface. This resulted in a

Bronze "magic" mirror, Han dynasty. The pattern cast in relief on the back of the mirror is reflected upon the wall by the polished front surface. (Courtesy of the Shanghai Museum, People's Republic of China. Photo Asian Art Museum of San Francisco)

bright delineation of the pattern on the screen. This "magical" property, inherent in the manufacturing process, appeared in mirrors as early as the Han dynasty.

The mirror of ancient China, aside from its conventional uses and supposed magical properties, represented lofty concepts of the cosmos. Because its physical reflection was somehow analogous to mental reflection, the mirror became the philosopher's touchstone for incisive metaphor. The sages of ancient China, who believed that in the process of metaphysical reflection true wisdom was obtained, used the mirror as a metaphor to characterize the kind of mental reflection required of a wise man. One of the earliest uses of the mirror metaphor in this regard is attributed to Chuang Tzu, a philosopher who flourished about 350 B.C. He described sagacity and nobility as follows: "When water is tranquil, its clearness reflects even the beard and eyebrows. . . .

The Mirror and Man

If water is clear when it is tranquil, how much more so is the spirit? When the mind of the sage is tranquil, it becomes the mirror of the universe and reflection of all things."[25]

Chuang Tzu also dealt with "calmness of mind." He stated that the mind of the perfect man is like a mirror. It does not lean forward or backward in its response to things. It responds to things but concedes nothing of its own. It is, therefore, able to deal with things realistically. Put another way, the mind of the sage lacks any anger itself. It is like a shining mirror in which a beautiful object produces a beautiful reflection, and an ugly object an ugly one. But the mirror itself has no likes or dislikes.[26]

A much more subtle mirror analogy was drawn by the philosopher Han-fei-tzi, in the third century B.C. It expresses the "Know thyself" philosophy. He wrote: "A mirror in which one does not see the flaws in one's face is like a tau [a philosophical method of reasoning] by which one is not enlightened of the wickedness of sin. If the eye has no mirror, it has no means of straightening up the hair on the temples and eyebrows; if man's self has no tau he has no means by which he knows his errors."[27]

It is clear that the concept of wisdom was embodied in the mirror. Wisdom obtained through undistorted reflection or contemplation was regarded as true and complete receptivity, for the reflecting power of the mirror typified the transcendental intelligence of heaven with its shining sun, which the mirror through its reflection could bring to earth.

Wisdom, the cosmos, and the phenomenon of reflection in mirrors were intertwined in the minds of Chinese philosophers. Their study of heaven, the phases of the moon, the rotation of the stars around the pole star, and the reflection of the moon and the stars in the lakes and rivers below gave rise to questions we still seek to answer today. What makes up the universe? One answer is provided in the strange and sophisticated philosophy expounded by the Hua-yen school of Buddhism in the seventh century. Its abstruse meaning is best explained by a unique mirrorlike arrangement known as

Indra's Net. Hua-yen taught that everything in the universe is manifested in infinite number and variety in every other thing; that is, each element of the universe contains within it all the other elements of the universe. This idea finds the unity of the universe in its multiplicity. This thought, no doubt, was incomprehensible when it was introduced, but it was explained by the structure and action of Indra's Net. In the hall of Indra, the chief Hindu atmospheric deity taken over by Buddhism to become its protective deity, was suspended a net, on each loop of which was hung a large lustrous pearl. On the surface of every pearl was reflected the image of all the other pearls, which themselves contained the reflected images of all the other pearls, so that on each pearl appeared an endless infinity of reflected images. The infinite reflections in any pearl visually represented the concept "that every individual thing in the universe is manifested in infinite number and variety in every other individual thing."[28]

Despite the explanation afforded by Indra's Net, it was a difficult philosophy for many students of Buddhism to understand. It remained for the great teacher Fa-tsing, 643–712, to explain it with a simple mirror demonstration. He took ten mirrors and arranged them, one each at the eight compass points and above and below, in such a way that they were about ten feet apart from one another, all facing inward. In the center he placed a brightly illuminated Buddhist figure, which was reflected innumerable times by the complex of mirrors; the one figure resulting in an infinity of reflected figures. Each mirror not only reflected the image of the other mirrors, but also all the images reflected in each of those other mirrors. When students saw the many figures produced by one, Fa-tsing was able to say that this represented the philosophy that the myriad are one, and the one is the myriad.[29] In their sum total, they constituted the Supreme Ultimate, yet each separate object also contained the Supreme Ultimate. For instance, considering the moon as an example of the Supreme Ultimate of which there is only one in the sky, it is reflected in a thousand lakes, rivers, and streams; yet from this it cannot be said that the moon itself is divided.[30]

Fa-tsing's mirror demonstration to explain an esoteric philosophy shows something of greater significance. The early Chinese must have had a good understanding of the optical properties of the mirror. For example, about one-hundred years before Fa-tsing, the scholar Lu Te-Ming remarked: "There is the mirror and the image, but there is also the image of the image; two mirrors reflect each other and images may be multiplied without end."[31] The Chinese understanding of the mirror, of course, dates back well over a thousand years before Fa-tsing. Curved mirrors were already understood then and incorporated in mirror design. Small mirrors were made with slightly convex reflecting surfaces to assure the reflection of a full face. A verification of this comes from an early encyclopedia which explained that in casting mirrors the ancients would give a small mirror a convex shape, "for all mirrors will reflect a man's face large if they are concave and small if they are convex. In a small mirror one cannot see the whole of a man's face, for which reason the surface is made slightly convex."[32]

The Chinese also knew as early as 1000 B.C. that the concave mirror was also a "burning mirror" able to start fires by concentrating sunlight. During the Chou period, around 700 B.C., the use of the burning mirror was common. The farmer and the hunter carried on the left side of his girdle a metal mirror to draw fire from the sun, and on the right, the borer for obtaining fire from wood. The mirror made fire during sunshine, the borer when it was cloudy. The burning mirror was called a *yang-sui* (sun igniter), and it is described in an ancient document as "made of bronze and . . . when held against the sun, it will provide fire, which is obtained through being caught by a heap of dried Artemesia leaves."[33]

Later in the Chou era the burning mirror became a cosmic symbol. It was used in certain governmental ceremonies to produce a "state fire" for sacrificial purposes. To be used on such occasions, the mirror itself had to be cast during certain ceremonies exactly at midnight and on the day of the solstice. A variation of this kind of mirror was used to "receive the brilliant water from the moon," apparently by exposing it during clear moonlight nights when it would be covered by the dew. The water thus obtained was used in sacrifices. Obviously, this parallelism to the concave mirror's ability to collect the sun's rays to start a fire did not hold for collecting water from the moon. It simply acted as a cup to collect the dew of night. This misconception, no doubt influenced by the awe felt toward that heavenly body, may be forgiven in view of the sophisticated knowledge the Chinese did possess.

An important optical property of the concave mirror was recognized by the Chinese as early as the sixth century B.C. By placing a lamp in front of a concave mirror, they devised a searchlight which they called *han kuang wa* (light-containing dishes). The light reflecting from a concave metallic mirror three feet in diameter was credited with lighting up a room at night. This was a remarkable discovery, for it was the forerunner of the first practical searchlight, which did not appear until 2,400 years later.

A school of thought called Mohism was established in the fourth century B.C. by the great philosopher Mo Ti. Many of his followers were artisans and engineers because he believed in a utilitarian approach to life. His hero was the legendary engineer Yu, who many centuries before had dug canals and built dams to protect North China from floods. Before the close of the fourth century B.C., a technical document appeared known now as the Mohist Canon, the *Mo Ching*. It revealed, among other scientific matters, an excellent perception of mirror optics. The document has definitive, though qualitative, statements about the properties of plane, concave, and convex mirrors. It appears to predate work done by the Greeks in this field. Certainly, there is no evidence to show any connection between the Greeks and the Chinese of that period.

A review of some of the statements in the *Mo Ching* as translated by Joseph Needham in his remarkable work *Science and Civilization in China* demonstrates the nature of the Chinese understanding of mirror optics. Starting simply with the plane mirror, the *Mo Ching* states that "standing on a plane mirror and looking downwards, one finds that one's image is inverted." And "a plane mirror has only one image. Its shape, bearing, colour, white or black; distance, far or near; and position, slanting or upright—all depend on the (position of the) (object or the) source of light. If now two plane mirrors are placed at an angle there will be two images. If the two mirrors are closed, or opened (as if on a hinge), the two images will reflect each other. The reflected images are all on the opposite side (from where the eye is)."[34]

The properties of the concave mirror were well understood by the Mohists: "With a concave mirror the image may be smaller and inverted or large and upright." This is something you, no doubt, have noticed any time you looked into the bowl of a soup spoon. From the distance your image is small and upside down; as you bring the spoon closer to your face, it will suddenly appear upright and enlarged. You may feel that such observation made by the ancient Chinese is quite ordinary, but reading their next statement you can better appreciate their knowledge of the concave mirror. Here they were dealing with a light source held in front of the mirror in searchlight fashion, and their observations were astute:

> (Take first) (an object in) the region between the mirror and the focal point. The nearer the object is to the focal point (and therefore the further away from the mirror), the weaker the intensity of light will be (if the object is a light source), but the larger the image will be. The further away the object is from the focal point (and therefore the nearer to the mirror), the stronger the intensity of light will be (if the object is a light source), but the smaller the image will be. In both cases the image will be upright. From the very edge of the central region (i.e., almost the focal point), and going toward the mirror, all images will be larger than the object, and upright.
>   (Take next) (an object in) the region outside the

center of curvature and away from the mirror. The nearer the object is to the center of curvature, the stronger the intensity of light will be (if the object is a light source), and the larger the image will be. The further away the object is from the center of curvature, the weaker the intensity of light will be (if the object is a light source), and the smaller the image will be. In both cases the image will be inverted.

(Take lastly) (an object in) the region at the center (i.e., the region between the focal point and the center of curvature). Here the image is larger than the object (and inverted).[35]

It is curious that the *Mo Ching* discusses all the possible positions the light source could have in front of the mirror except the focal point: the most efficient position for searchlight illumination. At the focal point the reflected beam is projected to its greatest distance, theoretically to infinity. It might be that the Chinese could not describe or understand the appearance of an image as large, small, or inverted at that point because the image loses its identity and appears only as a blob. Another mystery remains. With their understanding of the optical properties of the concave mirror, why did not the Mohists mention the burning mirror, an item used for utilitarian and ceremonial purposes as early as 1000 B.C.?

Regarding the optically simpler convex mirror, the *Mo Ching* has this to say: "With a convex mirror there is only one kind of image. The nearer the object is to the mirror, the stronger the intensity of light will be (if the object is a light source) and the larger the image will be. The further away the object is, the weaker the image will be. But in both cases the image will be upright. An image given by an object too far away becomes indistinct."[36]

To have developed this technical understanding of mirrors in the fourth century B.C. required the proficient use of an experimental method that was only made possible by the fine craftsmanship of early mirror makers who were able to make suitably curved or plane surfaces and to polish them smoothly and brightly. Unfortunately, these experimenters

The Mirror and Man

lacked the geometry needed to quantify their findings. It remained for the Greeks to carry out this crucial step which led to the full-fledged science of mirrors.

Beyond the Mohist studies there is no further evidence of any significant Chinese contribution to the science of mirrors. This is an anomalous situation in the development of a civilization where, contrariwise, science on this subject ended and its implications in art and religion grew and flourished for more than one thousand years thereafter.

The artistry and craftsmanship of the mirror reached a peak near the end of the eighth century. Then it declined and came to a virtual halt before the close of the Tang dynasty in 907. The causes, again, were political disorders and a subsequent decline of the economy that led to a shortage of copper and a drastic reduction in mirror production.

The level of earlier achievements was never again reached. Ensuing dynasties continued to make decorative bronze mirrors, but the only apparent innovation was the introduction of a handle that became very fashionable. Even that idea was the revival of an old one. Emperor Wu-Ti of the first century B.C., a great collector of foreign objects of art, introduced mirrors with handles that became popular among dancers of pantomime. They did not spread quickly among the people because of the deeply rooted attitude in favor of round cosmic mirrors, but eventually these disappeared and mirrors with handles came into vogue.

By the time of the Ming dynasty, 1368–1644, mirrors were made of poorer alloys and were of cruder workmanship. The artful fabrication techniques of the past were lost as a result of the earlier invasions and the periods of foreign rule around 1300 under the Mongols. In the Manchu dynasty, 1644–1912, the deterioration was complete. There remained only cheap brass "lucky" mirrors made as personal charms. The metal mirror was supplanted by the importation of foreign glass mirrors that were soon produced very cheaply in China itself.

The rise and fall of the art of the mirror reflects the glory and the decline of the civilizations that have produced them. But the heritage of the Chinese devotion to the mirror, as influenced by Buddhist teachings in the pre-Tang and Tang eras, still lives on in Tibet.

In the seventh century A.D. Tibet forced itself on the attention of its Chinese neighbors. Its young king, Song-tsen Gampo, pressed on into China itself in 635 and demanded and eventually received a Chinese princess as his bride. To Song-tsen Gampo is attributed the introduction of Chinese culture among his people, including writing and the Buddhist faith. The Tibetan influence on China did not last long, but Buddhist influence from China and India established that religion as the predominant one of that country. Mirror rituals developed then still exist.

A ritual of bathing the image of Buddha, which was common among the Chinese Buddhists of the fourth century, is performed annually on the eighth day of the fourth moon and the fifteenth day of the ninth moon. It is to commemorate the birth and first bath of Buddha. On these dates the faithful express deep devotion to Buddha and bathe his image. For this they are promised the fulfillment of all their wishes.

This ceremony, modified by local customs and adopted by the Tibetan Buddhists, is still performed by the Tibetan church. A lama conducts the rite. He assembles the image of a Buddhist deity, a mirror, a flask filled with water, a small basin, and a piece of cloth. The person offering the devotion places the image on the altar table so that it faces the mirror. He places the basin and flask between the image and the mirror. Then he begins to chant the sixty stanzas of a liturgy. At specified intervals he pours water from the flask into the basin, holding the flask so that the emanations imagined as flowing from the image pass through the water to the mirror, from which they are reflected back to the flask. Then the lama dabs the mirror in four places with the piece of cloth. This act is repeated a number of times during the long recitation, after which the water is poured back into the flask. This symbolism dates back to the early Buddhist use of the mirror as a metaphor of the mind that becomes dirtied by worldly thoughts, just as a mirror collects dust. This act of washing the mirror while a deity is reflected in it cleanses the mind of the devotee of world debris, illuminates it, and provides a mystical union with Buddha.

In another interpretation, the mind is considered a two-sided mirror. On the one side the realistic mind (*manas*), reflects, with error, the external world. On the other side the superior discriminating mind (*buddhi*), reflects errorlessly because it is devoid of images. In the language of the two-sided mirror, the *buddhi* side of the mirror can represent the Buddha because it is void of competing images. In this sense the ritual of mirror washing in Tibetan Buddhism is the washing of the mind so that it can properly reflect the divine world in the form of a deity's body.

In Buddhist doctrine there is no Being, there is only a Becoming. The state of every individual is unstable, temporary, sure to pass away. It is only the union of mental qualities that makes the individual. The Buddhist disciple uses a mirror analogy to explain the doctrine of non-Being. "With recourse to a mirror, one sees the reflected image of one's face, but in reality this (reflection) is nothing at all. In the same way, with recourse to the personality aggregates, the idea of self is conceived, but in reality it is nothing at all, like a reflection of one's face. Without recourse to a mirror, one does not see the reflected image of one's face. Likewise, without recourse to the personality of aggregates one does not speak of a self."[37]

This relationship of the unreality of a mirror reflection with Being is reminiscent of the Prajnavanta Jataka equating the mirror reflection with the unreality of a dream.

There are many more mirror passages in Buddhist literature that explain the many abstractions of that philosophy. It is quite understandable, in this religion of meditation on one's self, to include the study of one's mirror reflection for self-understanding and to explain theistic thoughts in worldly ways.

NOTES AND SOURCES

1. E. Loubo-Lesnitchenko, "Imported Mirrors in the Minusinsk Basin," *Artibus Asiae* 35 (1973):26–27.

2. Friedrich Hirth, "Chinese Metallic Mirrors," Boas Anniversary Volume (New York: G. E. Stechert, 1906), p. 213.

3. Ibid., p. 226.

4. A. Bulling, "The Decoration of Some Mirrors of the Chou and Han Periods," *Artibus Asiae* 18 (1955):42.

5. Bernhard Karlgren, "Early Chinese Mirror Inscriptions," *Museum of Far Eastern Antiquities Bulletin* (Stockholm), no. 6, 1934, p. 16.

6. Ibid., p. 24.

7. Ibid., p. 40.

8. Alexander C. Soper, "The 'Jen Shou' Mirrors," *Artibus Asiae* 29 (1967):59.

9. "Character of a Beautiful Woman Grieving before Her Mirror," in *Fir Flower Tablets*, ed. Amy Lowell, trans. Florence Ayscough (Boston: Houghton Mifflin, 1921, 1949).

10. Schuyler Cammann, "The Lion and Grape Patterns on Chinese Bronze Mirrors," *Artibus Asiae* 16 (1953):272.

11. Ibid., p. 274.

12. Ibid., p. 275.

13. Ibid., p. 282.

14. Ibid., pp. 286–87.

15. Ibid., p. 288.

16. Kimpei Takeuchi. "Ancient Chinese Bronze Mirrors," *Burlington Magazine* (London) 19 (1911):318.

17. Hirth, "Chinese Metallic Mirrors," p. 231.

18. Ibid., p. 232.

19. Ibid.

20. Ibid., p. 219.

21. Jan Jakob Maria de Groot, *The Religious System of China*, 6 vols. (Leyden: E. J. Brill, 1910), 6:1001–3.

22. Ibid., p. 1001.

23. W. E. Ayrton and John Perry, "The Magic Mirror of Japan," *Proceedings of the London Society* 28 (Dec. 12, 1878):135–36.

24. Ibid., pp. 127–48.

25. Wing-Tsit Chan, *A Source Book on Chinese Philosophy* (Princeton: Princeton Univ. Press, 1963), p. 208.

26. Ibid., p. 207.

27. Hirth, "Chinese Metallic Mirrors," p. 226.

28. Derk Bodde, "The Philosophy of Chu Hsi," *Harvard Journal of Asiatic Studies* 7 (1942):18, n. 21.

29. Yu-Lan Feng, *A History of Chinese Philosophy*, tr. Derk Bodde (Princeton: Princeton Univ. Press, 1952), p. 353.

30. Ibid., p. 581.

31. Joseph Needham, *Science and Civilization in China*, 5 vols. (Cambridge: Cambridge Univ. Press, 1954–74), 4:93.

32. Hirth, "Chinese Metallic Mirrors," p. 224.

33. Ibid., p. 226.
34. Needham, *Science and Civilization*, 4:83.
35. Ibid., p. 84.
36. Ibid., p. 85.
37. Alex Wayman, "The Mirror as a Pan-Buddhist Metaphor-Simile," *History of Religions* 13 (1974):260.

Other sources for this chapter include: Whitney Allen, "Metal Mirrors of the Ancients," *International Studio* 85 (Sept. 1926):31–36; Patricia Berger, "Treasures from the Shanghai Museum: 6000 Years of Chinese Art" *Orientations* 14 (May 1983):36–49; Schuyler Cammann, "Chinese Mirrors and Chinese Civilization," *Archaeology* 2 (1949):114–20; Ardelia Hall, "The Early Significance of Chinese Mirrors," *Journal of the American Oriental Society* 55 (1935):182–89; A. Kaplan, "On the Origin of the TLV Mirror," *Revue des Arts Asiatique* 11 (1937):21–24; Oscar Karlbeck, "Notes on Some Early Chinese Bronze Mirrors," *China Journal of Science and Arts* 4, no. 1 (1926):3–9; Bernhard Karlgren, "Huai and Han," *Bulletin of the Museum of Far Eastern Antiquities* (Stockholm), no. 13, 1941; Albert Koop, *Early Chinese Bronzes* (London: Ernest Benn, 1924); F. D. Lessing, "Structure and Meaning of the Rite Called the Bath of Buddha," *Studia Serica Bernhard Karlgren Dedicata* (Copenhagen: E. Munksgaard, 1959); James M. Plumer, "The Chinese Bronze Mirror: Two Instruments in One," *Art Quarterly* 7 (1944):91–108; Milan Rupert and Oliver Todd, *Chinese Bronze Mirrors* (Peiping: San Yu Press, 1935): Alfred Salmony, "Chinese Metal Mirrors," *Hobbies: The Magazine of the Buffalo Museum of Science* 25 (1945):96–104; Ch'ung Wang, *Miscellaneous Essays*, vol. 2, tr. Alfred Forke (Leipzig: N.p., 1907–11); Alex Wayman, *The Buddhist Tantras* (New York: Samuel Weiser, 1973); Archibald Wenley, "A Chinese Sui Dynasty Mirror," *Artibus Asiae* 25 (1962):141–48; Perceval Yetts, *George Eumorfopoulous Collection Catalogue of the Chinese and Korean Bronzes* (London: Ernest Benn, 1929).

JAPAN

THE HISTORY of the Japanese mirror is rooted in that country's ancient Shinto religion, yet its early material design is of Chinese origin. Shinto does not have an ideology or a theology. Its rites are designed to *evoke* a sense of awe that inspires gratitude to the source and nature of being. A Shinto priest has stated that "living Shinto is not the following of some set down moral code, but a living in gratitude and awe amid the mystery of things."[1] With the development of the spirituality of Shinto in the centuries before the birth of Christ, there arose a need for some visible, tangible token of the presence of a god. This was the mirror of Amaterasu-omikami, the sun-goddess. Its origin is described in the following myth.

Once Amaterasu was so outraged by the adulterous behavior of the God-ruler of the netherworld that she shut herself into the Rock Cave of heaven, thus putting earth in darkness. Greatly disturbed by this, the many gods of Japan on the Plain of High Heaven came together to devise ways of luring the goddess from her hiding place. They lighted fires, danced and frolicked, recited liturgies, but to no avail. Finally they ordered two of the gods to make a mirror of metal taken from the sacred mountain. Then, tantalizingly, they spoke before the cave: "The august mirror in our hands is spotless and indescribably beautiful as though it was thine own august person. Pray open the cave door and behold it." Whereupon Amaterasu said to herself: "How is it that the

The Mirror and Man

Gods can enjoy such merry making even when the world is wrapt in darkness, since I concealed myself in this Cave?"[2] And so saying she slightly opened the cave door to peep out. At that moment the mirror was pushed through the opening, and when the goddess paused, seeing her brilliant reflection in it, a muscular God took hold of her and led her out of the cave to light up the earth again.

After this, the gods esteemed the mirror greatly for its powers. When Amaterasu decreed that her grandson Ninigi-no-Mikoto go down to the Great Eight Island Country, Nippon, the Cradle of the Sun, she gave him the mirror to be handed down to his descendants of the imperial line and said: "Regard the mirror exactly as if it were our August Spirit; reverence it as if reverencing us and rule the country with a pure luster such as radiates from its surface."[3]

In this myth the identification of the mirror with the reflection of the disembodied spirit of the sun-goddess was the terrestrial representation of the all-powerful sun. To the mirror were added the jewel and the sword, also gleaming things, as symbols of the lesser natural mysteries of the moon and the lightning flash. Later these came to symbolize the virtues of knowledge, benevolence, and courage, of which the mirror remained the dominant member. These three sacred treasures have been regarded as the emblems of the imperial throne of the Yamato court from earliest times. As the imperial family extended its authority, its ancestor, the sun-goddess, became the chief deity of the entire land and subsequently her sun mirror was enshrined about 5 B.C. at Ise. This was in keeping with Amaterasu's request when she sent her grandson to earth. This was passed down orally until writing was introduced in Japan by the Chinese early in the sixth century.

Having deputized the two gods Ame-no-Kayane and Futo-dama, who were the ancestors of the great national sorcerors, to accompany him, she commanded them, "Take care, both of you, of this mirror and guard it well."[4] This mirror is still preserved at Ise with the greatest care and dignity. In the eyes of the Japanese, at least until the close of

World War II, Ise had the same importance as has the Holy Sepulcher for all Bible-believing Christians.

This mirror, which is believed to be about eight inches in diameter and most likely of bronze, is kept in an enclosure that rests on a low stand covered with a piece of white silk. David Murray, in his book *Japan*, notes that "it is wrapped in a bag of brocade, which is never opened or renewed, but when it begins to fall to pieces from age another bag is put on, so that the actual covering consists of many layers. Over the whole is placed a sort of wooden cage, with ornaments said to be of pure gold, over which again is thrown a cloth of coarse silk falling to the floor on all sides."[5]

Amaterasu's mirror is always hidden from view, but in front of the shrine there is another mirror, unornamented and round, mounted on a wooden stand carved with a pattern of clouds and water, the symbols of purity. This mirror, about nine inches in diameter, is known as the Unsui. Like the mirror of Chinese tradition, it is able to ward off evil spirits or reveal any evil shadows that may be hovering near.

For many centuries the mirror and those who were skilled in mirror making have been revered. It is said that when the sun-goddess sent her grandson to create order on the islands of Japan, she included in his retinue a noble company of mirror makers. Thus these craftsmen were endowed with a heavenly ancestor of their own who had forged mirrors on the Plains of High Heaven in Amaterasu's likeness. This traditional endowment was made official by the compiler of the historical document the *Kojiki* (Records of Ancient Matters) in A.D. 712.[6] Because of this distinction, mirror makers were considered part of the imperial court. They were men of consequence comprising a trinity of privileged craftsmen—designer, molder, and polisher—whose preeminence carried on into the nineteenth century.

The *Kojiki* was compiled at the instigation of the emperor Temmu. The emperor's action preserved the tenets of early Shintoism, but the faith was modified by Buddhist teachings that included new attitudes toward the mirror. By the ninth century the Tendai sect of Buddhism flourished in

Japan. Mirrors were part of the rituals in the Tendai temples, and worshippers looked on the mirror with devotion. The Buddhist deity Emma-O, chief of the Ten Regents of Hades, derived from the Hindu god Yama Raja, who sat in judgment over the dead, was seldom pictured without the "soul-reflecting mirror" (*Hau-no-Kagami*).[7] It revealed to Emma-O the truth in the heart of the devotee and was instrumental in driving out evil. This was done by a curious alteration of the reflection-soul concept. Breath was considered the soul in the body, and in breathing on a mirror the soul of the devotee was spread before Emma-O, who relentlessly searched out the smallest parts.

Like the Chinese, the Japanese Buddhists believed the mirror to possess unusual powers, as revealed by a number of entries in the diary of the monk Ennin Daishu (793–864), during his travels in China between the years 838 and 847:

> 839: Fifth Moon, 2nd day—At sunset on board the ship we worshipped the deities of heaven and earth and made offerings of official and private silks, tie-dye cloth and mirrors to the Great God of Sumerjoshi which is on board ship.
>
> 847: Ninth Moon, 8th day—We heard bad news (probably information regarding the presence of pirates or some other hostile force) and were extremely frightened, but since there was no wind we could not start. The group on board cast away mirrors and the like in sacrifices to the spirits to obtain a wind.[8]

Ennin Daishu's experience and the powers of Emma-O are but two examples of the Japanese Buddhists' attitude toward the mirror. During the many years when Buddhism predominated in Japan and became strongly enmeshed within the Shinto religion, the mirror appeared in many pious guises. A mirror with the image of Buddha engraved on the reflecting surface was regarded as an embodiment of Buddha. Some images of Buddha had mirrors placed in the halo, and sometimes a mirror was kept inside the image, as in China. In ceremonies performed to purify the land where a Buddhist temple was to be erected, sutras, mirrors, and other emblems were buried as a means of purification. Such mirrors have been recovered from the foundations of temples later excavated. Mirrors were also used as votive offerings at Buddhist temples. This was substantiated by a remarkable discovery made in 1914 when a large number of beautifully preserved copper mirrors were recovered from the bottom of a pond adjoining the Dewa Shrine on Mount Hogurosan.[9] The purity of the mountain water had protected these metal mirrors from corrosion, making them easy to identify as tenth-century objects. Apparently the mirrors were thrown into the pond as votive offerings by women visiting the shrine.

During the latter half of the nineteenth century there was a resurgence of the pure Shinto religion and the emperor was restored in 1867. Four years later Buddhism was disestablished; the shrines and temples were purged of Buddhist emblems, the old Shinto practices and rituals were revived, and Shinto was established as the religion of state. Buddhism, however, did not completely disappear.

Ancient Shinto mirror rituals reappeared. The ancient act of prostrating oneself before a mirror while watching one's reflection was reestablished. The ensuing contemplation was done with a view of searching out the innermost secrets of the heart and conscience while being aware that those whom the worshipper had been taught to venerate were beholding him, and judging him by their standards of right and wrong, and in some mysterious way were grieving over or exulting in his daily activities. It was a soul-cleansing act buried in an obscure tradition that predates the mythical era of the more popular Amaterasu.

The earliest god Izanagi and his mate Izanami stirred the sea with a spear and created an islet. Descending upon it, they created Amaterasu and many other deities as well as the other islands of Japan. When Izanagi and his wife decided to return to the celestial regions, he called his children together and asked them to listen attentively to his last wishes. He then presented them with a disk of polished silver, bidding them each morning to kneel before it and there see reflected on their

The Mirror and Man

faces the impress of any evil passions deliberately indulged, and again each night to examine themselves carefully "that their last thoughts might be after the happiness of that higher world whither their parents had preceded them."[10] Their children faithfully followed these instructions, which were devoutly carried on by their descendants, erecting an altar of wood to receive the sacred mirror and keeping fresh flowers in special vases renewed continually upon it. "As a reward for their obedience and devotion they became the spirit of good, the undying Kami."[11] It is likely that the Izanagi myth was created not only to explain the presence of a righteous Amaterasu but to personalize the worshipper's link to his illustrious forbears by leading a virtuous life in their honor.

With the revival of Shintoism, Amaterasu was honored in a festival held every new year. A feature of this festival was the baking of *kagami mochi*, or mirror cakes. These mirror cakes were round and flat and arranged two together, one over the other, as food offerings to the sun-goddess and her coterie of lesser deities. Of symbolic importance, they were distributed annually as New Year's gifts to the living as well as offerings to the departed to remind the people of the origin of the unbroken dynasty of rulers. The royal extension of this reminder was a reproduction of the sacred mirror at Ise enshrined in the sanctuary of the Imperial Palace in Tokyo. This divine object represented the spirit of the Imperial Ancestress, before which the emperor did daily obeisance. After World War II, Shintoism appeared doomed when General Douglas MacArthur ordered the separation of the state from the religion. But the Shinto priests turned to the people, and there followed an overwhelming wave of feeling for their native religion. Popular support was overwhelming; more than fifty million people contributed to the rebuilding of the Ise shrines in 1953, and the mirror of Amaterasu still lives.

In 1965 an unusual religious mirror was acquired by the Oxford Museum of the History of Science as a result of the bequest by a Henry Minn of Oxford and Cassington.[12] It appears truly magical; for, while the mirror has the usual convex surface and column of Chinese characters in relief on the back, it casts an image of a Buddha instead of the expected Chinese pattern when held up to the sunlight. The explanation lies in the fact that the back of the mirror consists of two layers: the Buddha in relief on the inner layer, responsible for the reflected image, is concealed by the outer false back (fig. 7). The inscription on the false back reads: "Adoration for Amida Buddha." The inscription is a mantra, or invocation, still used by the Pure Land School of Buddhism in China and Japan. The school had its origin in China as the Fellowship of the White Lotus in A.D. 402 and was founded as the Pure Land School in twelfth-century Japan.[13] The deity invoked and made visible is the Japanese Amida Bosatsu, the personification of compassion, the Buddha of Infinite Light, who assures the faithful rebirth in the "Pure Land in the West."

The composite mirror is clearly of a reverential nature combining the mantra (invocation), mandala (magic circle), asana and mudra (ritual gestures), and yantra (symbolic diagrams conceived in meditation represented by the reflection). But more important, this mirror casts further light on the Buddhist penchant to use mirror symbolism to express or propagate its philosophy. By exploiting an accident of manufacture, some wise priest created, by a bit of benign trickery, an arcane effect: making visible through the mystical power of the sun an apparitionlike image of the Buddha of Infinite Light.

As for the mirrors themselves, archaeological evidence from more than one thousand ancient bronze mirrors found in Japan since 1912, mostly in burial sites, suggests that the metal culture of Asia came to Japan about the third century B.C.[14] Simple mirrors to explain the one first enshrined in Ise were made about that time. By the first century B.C. Chinese culture had spread into the Korean peninsula, and Han mirrors were among the precious things the Japanese brought from Korea in the early days of sea traffic.

The earliest document dealing with the direct contact of the two nations is the official Chinese chronicle *The History of the Three Kingdoms*, written in the fifth century. It is still

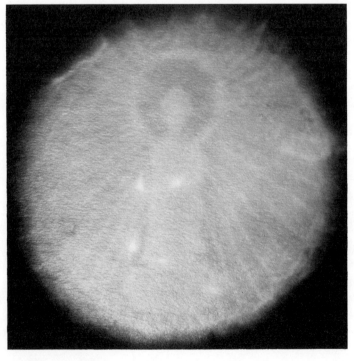

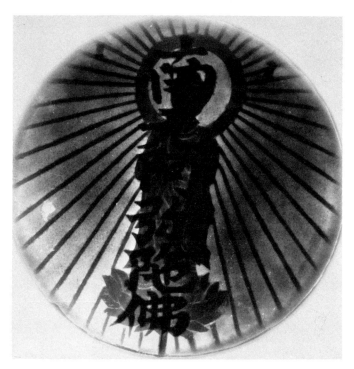

Composite mirror. Top left: *the back, showing the six characters cast in relief;* bottom left: *the reflection of sunlight onto a white screen, showing the figure of Amida Buddha;* above: *X-ray revealing the hidden relief responsible for the light pattern shown above. (Courtesy of the Museum of the History of Science, Oxford)*

regarded as a reliable source of information about ancient Japan. The *History* relates that in A.D. 238 ambassadors arrived from the empress Miyako, ruler of one of several little kingdoms of present-day Japan, with gifts for the Chinese court, and in January of the following year the emperor Ming-ti ordered a number of presents to be given to the ambassadors, which included silks, gold, pearls, and 100 bronze mirrors. In A.D. 240 another mission came to Japan, and mirrors were given again.[15] These tributes were continued from time to time, and in this way a great number of artistic bronze mirrors of the Han period found their way into Japan. The artisans of that country soon produced excellent mirrors of their own, patterned after the Chinese style of

The Mirror and Man

the middle of the Han period. The Kinai area, in the present vicinity of Nara, Kyoto, and Osaka, became a mirror-production center, producing mirrors affordable only by the nobility, who displayed them as symbols of prestige or objects of veneration.

The Chinese influence on mirror design and manufacture reached its height during the fifth and sixth centuries; the Japanese, however, introduced some novel variations. Maintaining the basic convex reflecting surface, they fabricated the bronze bell mirror (*suzukagami*), which had from four to eight rigid bells attached to the rim. Some were made with blossom petals, and some of the backs, instead of having a molded design with a pierced boss, were decorated with cloisonné, pearl inlay, enamel, or engraved silver and gold. The mirrors range from two inches to three feet in diameter and were kept in cloths or wooden nests to protect them from tarnishing.

There was a marriage mirror that varied from the Chinese version both in exterior decoration and intrinsic meaning. It was decorated with the Japanese symbols of long life and happiness—the crane, the tortoise, the pine, and the bamboo—and the intent was different. It was given to the Japanese bride by her mother at the wedding ceremony to serve as a sentimental reminder of the mother from whom the girl would be separated and as an emblem of her pure soul. Proverbially, the mirror was the soul of a woman just as the sword was the soul of the samurai, and when the mirror was dim the soul was unclean. This symbolism helped maintain the Japanese spirit in a woman's life, so that for her the mirror, clear and bright, was an intimate friend to whom she could confess her innermost feelings and cleanse her soul. With such properties, it is not surprising that the marriage mirror became an heirloom, filled with ancestral spirits with whom the owner could commune and hopefully look for their appearance in its surface.

Chinese influence waned in some of the arts after the sixth century, and Japanese culture finally flowered during its Heian period, from 794 to 1185. Bronze mirrors of Japa-

nese form called *wakyo* were developed as distinct from the
mirrors of Chinese style known as *kagami*. The Fujiwara
mirrors, *wakyos*, of that period are the most elegant of all.
They bear graceful and naturalistic designs of Japanese creation and are less elaborate and heavy than Chinese mirrors.
Mirrors continued to be precious personal possessions more
desirable than jewels. They were kept in beautifully lac-
quered boxes lined with brocade and wrapped in silk as befit-
ted such a valued possession. A handle was added to the
mirror after the fifteenth century, and by the seventeenth
century, the beginning of the Tokugawa period, 1603–1867,
the mirror evolved in quite a new guise. In conformance with
that sophisticated age, mirrors became more popular, and
secular uses predominated. Mirrors for dressing and coiffing
appeared in theatrical dressing rooms and in geisha houses as
well as in the homes of the affluent. They were elaborately
decorated and became part of toilet sets, resting in drawers of
elegant dressing chests containing combs and other toilet articles. Personal appearance was most important, and men and
women strolling along the boulevards were seen peering into
their sleeve mirrors to check the lay of their brows and the
fall of their locks. As mirrors became an object of daily use,
reaching down to the peasant, people became less concerned
with their souls than with facial beauty.

The art of metal mirror making persisted until the nineteenth century, when Japan opened its doors to the world
and commercial methods of production were introduced;
glass appeared and took over. The trinity of mirror makers—
the molder, the designer, and the polisher—had to abandon
their efforts and turn their skills in other directions.

NOTES AND SOURCES

1. Mikiso Hane, *Japan—a Historical Survey* (New York: Scribners, 1972), p. 23.

2. Kogoshui, *Gleanings from Ancient Stories*, tr. Kato, Ginchi,
and Hoshino (London: Curzon Press, 1972), p. 22.

3. Doris M. Roger, "The Divine Mirror of Japan," *Asia* 36
(1936):651.

The Mirror and Man

4. James Hastings, ed., *Encyclopedia of Religion and Ethics*, 12 vols. (New York: Scribners, 1924–27), 8:803.

5. David Murray, *Japan* (New York: Putnam's, 1894), p. 308.

6. Roger, "Divine Mirror of Japan," p. 653.

7. Ibid., p. 652.

8. Ennin, *Diary: The Record of a Pilgrimage to China*, tr. Edwin O. Reischauer (New York: Ronald Press, 1955), pp. 122, 402.

9. Roger, "Divine Mirror of Japan," p. 650.

10. Charlotte M. Salwey, "Japanese Monographs," *Imperial and Asiatic Quarterly Review* 8 (1899):402.

11. Ibid.

12. Gerard L'E. Turner, "A Magic Mirror of Buddhist Significance," *Oriental Art* 12, no. 2 (1966):94–98.

13. Ibid., p. 97.

14. Umehara Sueji, "Ancient Mirrors and Their Relationship to Early Japanese Culture," *Acta Asiatica* 4 (1963):71–72.

15. Friedrich Hirth, "Chinese Metallic Mirrors," Boas Anniversary Volume (New York: G. E. Stechert, 1906), p. 221.

Other sources for this chapter include: William Aston, *Shinto* (London: Longmans, Green, 1905); Jean Herbert, *Shinto* (New York: Stern and Day, 1967); Shoki Nihon, *Nihongi-Chronicles of Japan from the Earliest Times to AD 697*, tr. William Aston (London: Allen and Unwin, 1956); Doris Roger, "Japanese Metal Mirrors," *Connoisseur* 100 (Aug. 1937):65–68.

PRE-COLUMBIAN
AMERICA

THE BLACK MIRROR of Dr. John Dee came from the New World. It was a prized possession of the Aztecs, who used it for divination. And it was an important accessory of their god Tezcatlipoca, the omnipotent scryer. As novel as this obsidian mirror may have been to Dee, it was a well-known item among Mexican Indians during the declining years of an established mirror tradition nearly three thousand years old. This tradition originated with the Olmec Indians of Mexico about 1200 B.C. and subsequently spread through Mesoamerica and parts of South America. The Olmec culture, which was the forerunner of the Mesoamerican civilization, sowing the seeds of a number system, the calendar, and hieroglyphic writing, became adept at making mirrors from three different iron oxide ores (hematite, magnetite, ilmenite) and iron pyrites (fool's gold). These materials were readily available in their domain, which lay within the modern Mexican states of Veracruz, Oaxaca, and Guerrero.

The earliest extant Olmec mirrors are made of magnetite. They were discovered in 1966 by Kent Flannery in early formative deposits at San José Mogote in the valley of Oaxaca. These mirrors were made between 1400 and 1000 B.C. Nearly identical mirrors were excavated by Michael Coe the following year in the early formative San-Lorenzo-phase deposits in San Lorenzo, Veracruz. These mirrors, crudely circular, about five-eights of an inch in diameter, are flat and polished only on one side. This suggests that they were used as inlays in ornaments of some kind. These mirrors, however, vanished sometime before 800 B.C., when mirror production there seems to have come to an end. This kind of mirror was apparently no longer made after 800. Another type appeared around a new center of Olmec culture at La Venta, about twenty-five miles east of the gulf port of Coatzacoalcos. The small, flat magnetite mirrors were succeeded by much larger concave mirrors made of ilmenite and hematite. This remarkable discovery was made by Philip Drucker, an archaeologist, and his team during their 1942–43 expedition.[2]

Seven oval concave mirrors were unearthed in the ceremonial ruins of La Venta; the largest is about 4¾ inches on the long axis and 3½ inches on the short axis, about ⅜ of an inch thick, and it weighs about twelve ounces. These mirrors are perforated along one edge, presumably for suspension. This appears to be verified by a jade figurine found at La Venta wearing what most likely is an oval concave mirror on the chest. (fig. 8). An altar found in the same excavation bears a relief sculpture of figures wearing these concave mirrors.

Sixteen similar concave mirrors have been found within the Olmec domain. They were made between 1200 and 800 B.C. The three iron-oxide ores used, which are rocklike substances, yield nontarnishing reflecting surfaces, and these mirrors still maintain their high polish. The concavities of these mirrors appear to be spherical, but upon closer examination nearly all of them have a parabolic form, a mirror shape most suitable for concentrating sunlight. Yet there seems to be a design flaw—or is it? The lateral and longitudinal lines of the oval mirrors have dissimilar parabolic curvatures, providing the mirrors with two focal points. Was this design accidental or purposeful? It is easier to make a round concave mirror that has only one focal point rather than an oval one with two focal points.

Jonas Gullberg, who studied the workmanship and opti-

Olmec jade figurine with iron-ore mirror affixed in the pectoral position. (National Museum of Anthropology of Mexico. Photo courtesy of John B. Carlson)

cal characteristics of the seven original La Venta mirrors, recorded the two focal points of each mirror and concluded that the mirrors must have been prominent items in Olmec culture. He states, "They have a gracefulness, dignity, and perfection that make it hard to think of them as incidental or even only ornamental. The concave side has received a care that would seem to go beyond the standards of even superb lapidaries."[3]

Setting aside the loving care shown by these ancient craftsmen toward their work, it is difficult to believe that a culture of 1000 B.C., without benefit of metal technology and tools, could produce a parabolic concavity in slabs of iron-ore oxides. Gullberg suggests that special grinding and polishing tools of wood may have been used in a sophisticated manner. Grinding and polishing materials were readily available. The iron-ore oxides used by the Olmecs were no harder than glass and could be ground with common sand and polished with rouge, which is a finely powdered ferric oxide.

John Carlson reviewed this speculation and refuted the

notion of sophistication by demonstrating that it was quite simple for the Olmec craftsmen to produce the parabolic form. Using the technique of the amateur telescope mirror maker, Carlson duplicated what he believes was the Olmec mirror-making technique, one that certainly would be more consistent with the limited technology of the time.[4]

All that is required are two flat pieces of the mineral from which the mirror is to be made. The top slab, called the work piece, is rubbed back and forth over the bottom one, called the tool piece, with a suitable grinding compound between the two. As this action proceeds the tool piece becomes convex and the work piece becomes concave. If this reciprocal hand stroke is continued while occasionally rotating the two pieces in opposite directions, a spherical concave mirror will result. By the use of very long circular grinding strokes, more material is ground out of the center of the work piece and the spherical curve is changed to a parabolic one. This will produce round mirrors. To make an oval Olmec mirror, one starts with a predetermined oval tool piece smaller than the work piece. The tool dimensions dictate the shape and size of the final concavity. The grinding motion is oval, and the pieces are not rotated relative to each other. With this method Carlson produced a mirror indistinguishable from an Olmec original.

The prominence of concave mirrors at the La Venta site and their appearance as chest pendants on figurines of noble appearance suggest a religious tradition based on the power of the mirror to bring the sun via reflection close to the beholder. Its concavity could capture the rays of the sun and create fire. This could easily have instituted a supernatural relationship between the sun, which was worshipped, and the mirror, making the latter the earth symbol of the supreme power. Variations of this sun-mirror connection exist in other ancient civilizations: the mirror shape of Egypt, the sacred fire from the sun for ceremonial occasions in early China, and the sun-goddess myth of Japan. This mirror, because of its miraculous alliance with the all-powerful sun, became the insignia of the priesthood and royal lineage that emerged and was worn as a display of divine power by the privileged. The inculcation of this notion was profound.

The Olmec civilization started to decline about the fifth century B.C., with a gradual disappearance of their culture over the following three to four centuries as the Mayan civilization emerged. The Mayas reached high cultural development about the third century A.D., and this continued for several hundred years. About the eleventh century the Mayan civilization collapsed with the rise of the Toltecs, who were succeeded by the Aztecs about A.D. 1300. This civilization was dominated by the god Tezcatlipoca of the Smoking Mirror. Throughout these changing cultures, and changing mirror materials from iron-ore oxides to iron pyrites and obsidian, the Olmec mirror tradition remained basically unchanged. But a smoking-mirror deity associated with a sacred heritage of royal lineage had evolved.

Carlson offers a highly plausible theory and supporting evidence for this idea, based on four intermeshing elements. The first two, alluded to above, are: (1) the concave mirror is designated a symbol of the sun and exploited to provide the ascendency of the elite. (2) It also has the supernatural power to create fire from the sun.

Carlson further expands the scope of his second element by relating the smoking mirror to Tezcatlipoca and rulership. He theorizes that smoke may have been produced more often than fire because many of the Olmec concave mirrors, as evidenced by the ones found, were too small to produce fire when directing sunlight upon tinder, but most certainly they did make smoke. This mirror-produced smoke from the sun, as it were, must have been considered awesome, for it eventually led to the all-powerful Aztec god Tezcatlipoca, which literally means "smoking mirror." This god of many faces ruled the earth's surface. He was the black martial god of the North, the spirit of witchcraft and black magic. He was the omnipotent scryer, who, looking into his smoking mirror of obsidian, divined and controlled the deeds of men. Both he and his father, Xuihticuhtli, the old fire-god, bestowed rulership and were associated with the succession of the royal

The Mirror and Man

lineage. The Toltecs, the predecessors of the Aztecs, also had a Tezcatlipoca, and the earlier Mayas had a God K, the Lord of Nine, or Infinite Generations, who has been identified with the black god. Figures of God K found in the Temple of the Inscription at Palenque show him with a mirror and a smoking tube implanted in his forehead, a clear forefunner of Tezcatlipoca. Here inscriptional evidence links God K as the protector of royal lineage and ancestry. In support of this there is a drawing on the sarcophagus lid showing profiles of the nobleman Lord Pacal of Palenque and God K, each having smoke-issuing mirror infixes on their foreheads. The connection is unmistakable.

The third element of Carlson's idea involves the traditional use of mirrors in scrying. In this case with the aid of the Tezcatlipoca figure, Mesoamerican rulership was bound up with priestly divinatory functions.

The fourth element of Mesoamerican association of the mirror and rulership comes from the Mayas in a specific way. In the Palenque hieroglyphic inscriptions in the context of an "event glyph," there is a hand holding a concave mirror. This glyph also appears on the Maya sun-god, Itzamma and some lesser deities. To interpret this glyph, Carlson refers to an unpublished earlier study by Jeffrey Miller in 1974 known as the "nen"/mirror hypothesis. "Nen is the Yucatec word for mirror and is so defined in the Yucatec dictionary. However, the entry following 'nen' is the phrase 'U nen cab,' literally, the people's mirror, which is defined as: The priest, chief, or governor of the land or village who is the mirror in which all the people see themselves." This connects the mirror figure with the leader in whom the people should see themselves reflected. Further linguistic and iconographic evidence provided by Miller relates the mirror to rulership. He believes "there was probably a 'mirror ceremony' involved in the transfer of royal lineal powers, heir designation, or accession to leadership."[5]

Convincing support for this belief can be found in a description by Fray Diego Duran in 1581 of the ceremony electing Montezuma II ruler of the Aztecs. The principal elector, Nezahualpilli, speaking to the gathering of the great lords of Mexico, proclaimed, "With your vote and consent we are to choose the luminary that is to give us light like one of the sun's rays. We are to choose a mirror in which we will all be reflected . . . a prince who will rule over the Aztec nation."[6] Montezuma was elected and enthroned. Then Nezahualpilli addressed him most gloriously: "O you most powerful of all the kings of the earth! The clouds have been dispelled and the darkness in which we lived has fled. The sun has appeared and the light of dawn shines upon us after the darkness which has been caused by the death of the king. The torch which is to illuminate Mexico has been lighted and today we have been given a mirror to look into."[7] Joyce Strauss notes that nowhere is there documented a more definitive statement concerning illumination, mirrors, and accession to rulership.[8]

The Olmecs also made mirrors of iron pyrite, an excellent natural material for the purpose. When the nodules are ground and polished, they make a brilliant white or near white reflecting surface. Since pyrite is a crystal and has a natural cleavage with flat surfaces and straight edges, small, well-defined brilliant flakes are easily removed. Because these polished pieces were quite small, they were assembled and cemented as tightly fitted mosaics on a backing of slate, ceramic, wood, or shell.

Pyrite mirrors were manufactured by the more cultured natives of Mesoamerica and South America and were disseminated through trade quite extensively. Mirrors of marcasite, another form of iron pyrite, were also made and traded. Examples of this kind of mirror have been found not only in Mexico but also in Guatemala, British Honduras, Costa Rica, Panama, Ecuador, and Peru. There is very little evidence of mirror use among Indian cultures north of Mexico. There are isolated cases: Marcasite mirrors among the Hohokams in Arizona, mirrors of mica among the Adena Indians of the Ohio area, and slate mirrors among the Tsimshian Indians of British Columbia.

In Mesoamerica, central Guatemala seems to have been

The Mirror and Man

a center of manufacture for pyrite mirrors. Circular mirrors varying in size from 3¾ inches to 6 inches in diameter with a thickness of about ⅛ inch have been found in ancient villages in that area. The craftsmanship of these mirrors is excellent. Generally, the mosiac elements are five or six sided and about ⅛ inch thick. All the edges are perfectly straight, and they make tight junctions with the sides of the adjacent polygons.

Very few pyrite mirrors of pre-Columbian origin have been found in South America, perhaps because of deterioration or lack of use. Even in the great civilization of Peru, which left evidence of magnificent temples laden with silver and gold, weapons and tools of bronze, and examples of excellent engineering feats, there is very little evidence of pyrite mirror manufacture or use. Yet the earliest documented foreign information (the Peruvians, including the Incas, had no written language) suggests such an industry.

In a 1526 account by Juan de Sáamanos, the representative of King Carlos of Spain who accompanied Pizarro to the New World, we find the first mention of Peruvian mirrors. Pizarro, quartered then on the Pacific coast of Colombia, ordered his pilot, Bartolomé Ruiz, to explore the coast south of his position. Ruiz sailed along the coast of Ecuador. He was the first European to cross the equator on the Pacific side, and he found increasing signs of culture. Finally, he accosted a heavily laden raftlike boat sailing northward from Peru, the Inca Empire, to trade goods with northern neighbors. This was the first meeting of Caucasians with natives of aboriginal Peru. Ruiz captured the vessel, the first Inca booty to fall into the hands of the Spaniards. With a great quantity of trading goods, from ornaments in gold to colorful costumes in wool and cotton, were silver-framed stone mirrors.[9] The mirrors were probably pyrite.

Examples of pre-Columbian Peruvian pyrite mirrors have been found in recent times. One of the earliest is dated between 400 B.C. and A.D. 100. It is encased in a frame of gilded copper and has a handle. The periphery of the mirror is surrounded with twenty copper birds (fig. 9). This mirror,

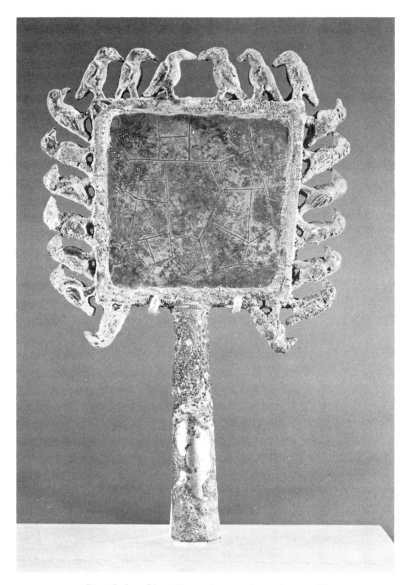

Pre-Columbian Peruvian pyrite mirror. (From
Precolumbian Art of South America, *by Alan C.*
Lapiner, *publisher, Harry N. Abrams, Inc.)*

The Mirror and Man

which is 8¼ inches high and 5⅛ inches wide, was fabricated by artisans of the Mochica culture, a forerunner of the Inca civilization. That mirrors were then in use is seen in the exquisite paintings on the magnificent pottery for which that culture was renowned. The detailed illustrations show women in fine dress with bobbed black hair, eyelashes plucked, cheeks rouged, and lips painted. Mascara darkens their eyelashes and brows. Such attention to facial features by these women surely required mirrors.

In his book *An Ancient World Preserved*, Frederic Engels reports finding in Peru a small piece of very shiny hard stone cemented to a clay tablet which he believes to be the earliest mirror, about 1500 B.C., extant from South America. There are some sketchy reports by others of Peruvian obsidian mirrors that are believed to have been made between 1250 and 850 B.C.[10]

Historically, the next bit of information comes in 1609 from the writings of Garcilaso de la Vega, whose mother was an Inca princess and whose father was Castilian. Garcilaso is considered the first authority on the civilization of the Incas and on the conquest of Peru. From his narrative it appears that mirrors had reached a high level of sophistication and were used for secular and religious purposes. He writes that ladies of royalty looked at themselves in silver mirrors, but ordinary women were not allowed to use silver and had to resort to common ones of bronze. Mirrors were for women only; men never used them because to do was considered shameful and effeminate. Further, Garcilaso attributes fire making with mirrors to the Incas of his time.[11] At the time of the great sacrifice of the feast of the sun (the state religion of the Incas was sun worship) new fire was made by the chief priest by means of a highly polished concave mirror that was part of his armlet. Twentieth-century confirmation of this religious rite comes from the ruins of Machu Picchu, a citadel of the Incas near their capital of Cuzco. This remarkable city of granite, built during the first half of the fifteenth century and discovered in 1911, contains a temple for the Virgins of the Sun, priestesses of high standing. In the burial

Pre-Columbian America

place of Mama-Cuna, mother superior of the Virgins of the Sun, was found a ceremonial bronze concave mirror. No doubt like their Mesoamerican neighbors, the Peruvians had a sun-mirror tradition, a tradition that very possibly arose from a single source, for there is strong evidence that there was contact between these civilizations. The numerical and calendric systems of the Peruvians, the Mayas, and the Mexicans were almost identical, and many features of their religions as well as their deities were similar.

Of the mirrors from pre-Columbian America, the Mexican mirror of obsidian is the most fascinating. The black glass, spewed as lava from smoking volcanoes, was easily identified with Tezcatlipoca, the black god, when the potentialities of this mineral became known. Its hardness and ease of flaking made it most suitable for piercing and cutting weapons such as arrow heads, spear tips, and knife blades. When worked in slab form and ground and polished, its high surface luster made it an effective reflector. Because of its easy availability in Mexico, obsidian became the dominant material for mirrors in the Toltec and Aztec civilizations. These mirrors were made in round or rectangular forms and sometimes framed in wood or precious metal. A particularly attractive obsidian mirror framed in a carved gilded wooden frame is now in the American Museum of Natural History. It is 10½ inches in diameter and ¼ inch thick. Its frame, covered with thin, dull gold leaf, is 1⅛ inches wide and 1 inch thick.

Obsidian itself became deified as Tezcatlipoca, and his idol was often made of that stone with mirrors in his forehead and on his missing right foot, which was supposedly torn off in a confrontation with the earth monster. As the supreme deity in the Aztec pantheon, Tezcatlipoca was feared by the general populace because of his power to mete out rewards and punishments according to an individual's thoughts and behavior as he divined them in his smoking mirror. Often he was provided with another attribute to strengthen his fearful image. This is best described in an eyewitness account of Captain Bernal Díaz del Castillo in

The Mirror and Man

1568. He was an aide to Cortez, the conqueror of Mexico. Invited to visit with Montezuma in the ruler's temple, Cortez asked to be shown the Aztec gods. The king took them into one of the sacred rooms where there were two highly adorned altars over which stood two high figures. "The one on the right was Huitzilopochtli, their war god with a great face and terrible eyes. On the left was the other great figure, with a countenance like a bear, and great shining eyes of the polished substance [obsidian] whereof their mirrors were made. These two deities, it was said, were brothers; the name of the last Tezcatepuca and he was the god of the infernal regions. He presided over the souls of men."[12]

Obsidian mirrors as the eyes of the gods were part of the Toltec and Aztec mirror tradition. One can imagine how effective they were when, with the idols facing the setting sun, the angry, blood-red orbs glared menacingly at the fearful worshippers. The obsidian eyes reflecting the light gave the idols an aura of watchfulness and life.

Mirror eyes were involved in ceremonies dealing with such disparate subjects as death, crop fertility, and chastity. When the Aztec ruler Tizoc died in 1486, his funeral rites embodied this tradition. Tizoc's body was placed before the idol of Huitzilopochtli to be cremated. Several black-painted attendants, naked except for loin cloths, were given sticks to rotate the body in the fire. Then there appeared a figure dressed as the king of the underworld. His demonic appearance was heightened by the two shining mirrors he wore as eyes. On each shoulder he wore a mask with mirror eyes. Similar masks appeared on each elbow and knee, and there was a mirror-eyed mask on his stomach. As he danced about the fire, the gleaming, all-seeing eyes made his appearance so frightful that few could look at him without shuddering. His function was to oversee every move of the attendants and to urge them to hurry the turning of the corpse in the flames.

In contrast to this grisly scene, there is an idyllic account of obsidian eyes by Juan de Tovar. It appears as one of the monthly ceremonies described in Tovar's illustrated Aztec calendar.[13] This secular calendar with a 365-day year had

eighteen months of twenty days each plus five additional
days, as distinct from the religious calendar, which had a
260-day year divided into twenty thirteen-day months. The
ceremony of the fourth month is named *Toxcatl*, which
means drought or sterility of the land. This was in April
when the land was arid and the rain was about to come.
Tovar described the ceremony (which incidentally includes a
clue to the conquering Spaniards' invincibility):

> Then they brought forth the arms and trophy of the
> great god [Huitzilopochtli], who was like Jupiter among
> the Romans. From the different seeds of the land they
> made dough and with it formed a face which was a
> great idol. . . . For eyes they gave it two mirrors which
> were always guarded in the temple, and which they
> called the eyes of god and thus when the ancients saw
> Spaniards with eyeglasses they said they had the eyes of
> their [war] god. At this time they put the face and arms
> of the idol before the people in order to signify that the
> all-powerful lord, whose insignia these were, possessed
> the power to give them a good year, and thus it was
> prayed for with great supplications and sacrifices all this
> month.[14]

The ceremony of the eleventh month, *Teotleco*, or the
advent of the gods, tells of an annual miracle Huitzilopochtli
performed this month to determine the faithfulness of a selected group of twenty men and twenty women who remained in the temple for one year to await his arrival. During this period they were required to carry on the duties of
the temple under special conditions: the men and women
were to remain isolated from each other and were to observe
a rigid law of seclusion and chastity. When they completed
their year-long service, they came out as honored or debased,
depending on whether they had been faithful in chastity.
This was known from the little mirrors called the eyes of god
which were kept concealed in an opulent container in a sacred corner of the temple. Though they were enclosed, the
eyes guarded the temple and watched all that took place

therein. The priests uncovered the eyes on the day of the advent of the god. If they were clear, it meant that the devotees of that year had conducted themselves well and they were dismissed with high honor and many gifts. But if the eyes were full of dust or hair, it was an infallible sign chastity had been violated. The dishonored then confessed their sin and were hanged because they had filled the eyes of the god with dirt and wickedness.

We can see from these and from other tales brought home by the Spaniards why New World mirrors were as much prized as articles of gold and silver. Evidence that such mirrors reached Europe is contained in the official inventories of the day. In the spoils sent by Cortez to the king in 1519, there is listed "a mirror placed in a piece of blue and red stone mosaic-work, with feather work stuck to it."[5] Another list from Diego de Soto included "a mirror with two faces; a mirror with a figure of guastica; . . . a round mirror like the sun; a mirror with the head of a lion; a mirror with the figure of an owl."[6] Although these lists do not describe Dee's black mirror, there is no doubt where it came from. It is no wonder that Dee, convinced of its supernatural attributes, held his Shew-Stone in such high esteem and valued the visions seen in it by his scryer, Kelly. Its heritage was truly mystical.

NOTES AND SOURCES

1. John B. Carlson, "Lodestone Compass: Chinese or Olmec Primacy," *Science* 189 (1975):754.

2. Phillip Drucker, R. F. Heizer, and R. J. Squier, "Excavations at La Venta, Tobasco, 1955," *Bureau of American Ethnology Bulletin*, no. 170, 1959, p. 181.

3. Ibid., p. 282.

4. John B. Carlson, "Olmec Concave Iron Ore Mirrors: Technology and Purpose," *XLII Congres International des Americanistes* (Paris), 1976, pp. 3–4.

5. Ibid., pp. 7–10.

6. Fray Diego Duran, *The Aztecs: The History of the Indies of New Spain*, tr. Doris Heyden and Fernando Horcasitas (New York: Orion Press, 1964), p. 220.

7. Ibid., p. 221.

8. Joyce R. Strauss, "A Mirror Tradition in Precolumbian Art," M.A. thesis, University of Denver, 1977, p. 124.

9. Thor Heyerdahl, *American Indians in the Pacific* (London: Allen and Unwin, 1952), pp. 516–17.

10. Frederic Engels, *An Ancient World Preserved* (New York: Crown Publishers, 1976), p. 105.

11. Garcilaso de la Vega, *Royal Commentaries of the Yncas* (Lisbon: Office of Peter Crasbeeck, 1609), p. 203.

12. Bernal Díaz del Castillo, *The True History of the Conquest of Mexico*, tr. Maurice Keatinge (London: Printed for J. Wright by J. Dean, 1800), p. 140.

13. George Kubler and Charles Gibson, "Tovar Calendar," *Memoirs of the Connecticut Academy of Arts and Sciences* 11 (Jan. 1951):25.

14. Ibid., pp. 29–30.

15 Marshall Saville, *The Wood Carver's Art in Ancient Mexico* (New York: Museum of the American Indian, Heye Foundation, 1925), p. 87.

16. Ibid.

Other sources for this chapter include: Cottie Burland, *Art and Life in Ancient Mexico* (Oxford: B. Cassirer, 1948); Cottie Burland and Werner Forman, *Feathered Serpent and Smoking Mirror* (London: Orbis Publishing, 1975); Alfonso Caso, *The Aztecs*, tr. Lowell Dunham (Norman: Univ. of Oklahoma Press, 1958); Michael D. Coe, *America's First Civilization* (New York: American Heritage, 1968); Hans Disselhoff and S. Linne, *The Art of Ancient America*, tr. A. Keep (New York: Crown Publishers, 1961); Phillip Drucker, "La Venta, Tabasco: A Study of Olmec Ceramics and Art," *Bureau of American Ethnology Bulletin*, no. 153 (1952); Gordon Ekholm, "The Archaeological Significance of Mirrors in the New World," *Attidel XL Congresso Internazionale Degliamericanisti* (Rome), 1972; Thomas Joyce, *Mexican Archaeology* (New York: Hacker Art Books, 1970); Alan C. Lapiner, *Pre-Columbian Art of South America* (New York: Harry N. Abrams, 1976); John A. Mason, *Ancient Civilizations of Peru* (Harmondsworth: Penguin Books, 1957); Frederick Peterson, *Ancient Mexico* (New York: Putnam, 1959); Lewis Spence, *Gods of Mexico* (London: T. F. Unwin,

1923); Julian Steward, ed., *Handbook of South American Indians*, vol. 3, *Bureau of American Ethnology Bulletin*, no. 143; George C. Vaillant, *Aztecs of Mexico* (Garden City, N.Y.: Doubleday, Doran, 1941); Alpheus Verrill and Ruth Verrill, *America's Ancient Civilizations* (New York: Putnam's, 1953).

ANCIENT AND
MEDIEVAL EUROPE
The Material Mirror

MIRRORS OF BRONZE appeared almost simultaneously in the ancient civilizations of Greece and Etruria (a region of Italy now known as Tuscany) during the sixth century B.C. , though there is fragmentary evidence that they were in use on the Greek mainland as early as 1400 B.C. Almost certainly they came from Egypt. One of these early mirrors was discovered in a tomb at Routsi, near Pylos. It is a hand mirror of polished bronze with a handle of ivory found at the waist of an unknown warrior, who also had a dagger by his side, an arrow between his legs, and a bronze cup standing near his head. Did the soldier have to go into battle well groomed? We will never know. We also do not know whether the mirror was made in Egypt for export or fabricated by a Greek artisan after an Egyptian design. Except for this and a few other isolated finds, nothing further is known about mirrors there until the sixth century B.C. However, the ancient Greeks, during that eight-hundred-year period, probably became adept in the casting and polishing of bronze hand mirrors and may have exported some of their wares to Italy, where the Etruscans settled about 800 B.C. Because the contribution of the Etruscans to the history of the mirror

The Mirror and Man

lasted only a few hundred years, it may be examined before considering the much more extensive use of mirrors in ancient Greece. The Etruscans had migrated from Asia Minor, attracted by the rich deposits of copper in Tuscany and iron on the island of Elba. They possessed a highly developed culture, and now, with a source of copper and iron, developed an excellent knowledge of metals and metalcraft by which they attained a leading position of power and well-being. This is seen in their bronze mirror production, which was at its peak from about the sixth to the third century B.C.

Many Etruscan hand mirrors still exist. Hundreds were found during the nineteenth century and described by Edward Gerhard and G. Karte.[1] Subsequently, with the discovery of nearly two-thousand more of these mirrors, much has been added to our knowledge of them. They are basically from two periods: the Archaic, which extended from the end of the sixth century to the early part of the fifth century B.C., and the (early) Classical period, from the fifth century to the beginning of the third century B.C. These mirrors provide an insight into Etruscan culture—their craftsmanship with metals, their engraving skills, their artistic creativity, and, through mirror inscriptions, their language. Typically, these mirrors were manufactured of bronze, approximately 86 percent copper and 13 percent tin, in a disk shape with a tapering tang over which was set a handle of bone or ivory. This style changed during the third century B.C. when it yielded to a mirror where the disk and handle were cast as a single piece. The handles were often shaped as knobs or animal heads. The mirrors varied from about four inches to ten and one-half inches in diameter, and because of their relatively small size their reflecting surfaces were made slightly convex to cover the full reflection of the face. The two main mirror-manufacturing centers were in the cities of Praeneste, modern Palestrina, and Caera, modern Cerveteri.

The engravings on the backs of these mirrors most frequently depicted scenes from Greek mythology. The Etruscans, through extensive trade with their neighbors, learned the Greek heroic legends and their fascinating gods and apparently appreciated the human frailities of these deities. This is apparent in the Etruscan engravers' representations of these themes with amusement and affection rather than reverence. Indeed, some legends are presented with greater flourish than those found on any extant Greek monument. The story of Oedipus and his family, episodes from the Calydonian boar hunt, the labors of Hercules, Admetus and Alcestis, and many other mythological favorites appear repeatedly.

Hercules appears to have been a favorite. Details of his adventures occur often in these mirror engravings. Usually he is clothed in a lionskin worn knotted by its front paws around his neck and hanging down his back, and he is armed with his club. He is often seen in heroic attitudes, but sometimes he appears in tender poses. Figure 10 shows Hercules, bearded and with his club, refreshing himself between his labors by taking a sip of milk from the breast of his mother, who is represented as Juno. This is an alteration of the Greek myth, which denotes Alimene as Hercules' mother. In the background an unbelieving attendant watches as Hercules' mother holds her nipple to the hero's eager mouth.

The engravings of the Archaic period are characterized by love of life. Single figures are often shown moving swiftly, either running or flying. Classical period scenes have an even freer style, generally consisting of from two to four figures intimately related by their gestures and glances. In addition to these heroic settings, there are pictures of everyday activities ranging from toilet scenes to courting scenes. Judging from the elaborate coiffures in these mirror engravings, hair dressing, especially for women, appears to have been a fine art.

The engravings on the mirrors from the Classical period indicate a diminishing dependence on Greek materials and the emergence of purely Etruscan themes. One such theme is soothsaying, an art for which the Etruscans were renowned. With this they won the admiration and confidence of their Roman rivals, who in a number of cases requested the counsel of Etruscan soothsayers. On a mirror now at the Vatican there is engraved a winged, bearded man, identified as Calchas the seer, in the act of examining the entrails of an animal. In his hand he holds the liver. A lung with the tracheae rests on the table. This was a method of augury in many cultures. Providing Calchas with wings may have been the engraver's way of trying to endow him with a touch of divinity. Another Etruscan theme is a fascination with blood and gore. This is portrayed on a mirror back by a scene of the mythological struggle between the Olympians and the giants. The engraving shows the goddess Minerva in combat with the giant Akrathe. The goddess has been aroused to such a fighting fever that she has violently torn off the right arm of her hapless foe; blood spurts profusely from the socket of his shoulder while Minerva wields the dismembered arm as a gruesome weapon.

Another aspect of Etruscan culture can be drawn from the inscriptions on these mirrors. Personages are named and something is added about the scene. This indicates that the ladies and gentlemen of the period could read, and it provides an invaluable aid to the study of the Etruscan language.

A revealing feature of Etruscan mirror art was the portrayal of bawdy scenes. One mirror back in the Gerhard collection shows Menelaus and the goddess Aphrodite in the nude. This trend may have been influenced by the Roman conquerers of Etruria during the third century B.C.. They staged performances of male and female mime players in the nude. These performances, somewhat pornographic, became the most popular feature at the spring flower festival, the Floralia.

Despite the depiction of a variety of male and female figures in Etruscan mirror designs, available evidence suggests that mirrors were used only by women. When presented in scenes of daily life, the person with the mirror is always feminine. More important, Etruscan mirrors, which have been found only in funerary surroundings, appear to have been associated solely with female burials. The mirror's exclusive appearance in burial sites shows a link with the superstitions of the past. Indeed, the Etruscans, like other people of that time, believed that the mirror held the soul of

Hercules sipping milk from his mother's breast. (Courtesy of E. Gerhard, Etruskische Spiegel, *vol. 5, pl. 60, de Gruyter, Berlin, rpt. 1974. Photo Library of Congress)*

the person it reflected and after death wanted the assurance that body and soul went to the next world together. This becomes evident from the Etruscan word for soul, *hinthial*, which also means "image reflected in a mirror."[2]

Although the Etruscans perfected their skills by taking advantage of what they learned about mirrors from early Greek exports, the Greeks in their turn made no use of Etruscan innovations in line engravings. During the period from 600 to 300 B.C., when both cultures were at their height of mirror production, Greek mirror designs are free of Etruscan influence. The only common factors are the use of bronze and the slight convexity of the reflecting surface of the hand mirrors.

The extensive use of mirrors by the Greeks in that period is shown in scenes glazed on vases and terracotta statuettes. A woman holds the mirror in one hand while attending to her toilet with the other; or she supports it on her knee to have both hands free; or an attendant holds it up to her. Sometimes a mirror is shown hanging by a nail on the wall. In 1867 the first specimen of a Greek mirror was discovered at Corinth, which was a chief manufacturing center for mirrors. Others have been found at Mycenae, at Italysos, at Vaphio, and in Crete. The earliest of these mirrors, dated from 550 B.C., are known as Argivo-Corinthean mirrors. They consist of disks with attached flat handles that are rounded at the bottom. The handle is decorated with scenes in low relief. Sometimes the handle is in the form of a human figure terminating in a suspension ring.

Elaborate hand mirrors were manufactured during the sixth century B.C. reflecting the sophisticated art of that time. A beautiful example can be found in the New York Metropolitan Museum of Art (fig. 11). The handle is a nude girl, apparently of Egyptian influence, standing on the back of a crouching lion. Her figure is delicate, with ornamentation that extends this delicacy. In her left hand she is apparently holding a pomegranate. She has a close-fitting necklace with a single pendant and a band strung with amulets. Her

Greek bronze mirror, 6th century B.C. *Handle in the form of a female figure standing on a recumbent lion. (The Metropolitan Museum of Art, Fletcher Fund, 1938. All rights reserved, The Metropolitan Museum of Art)*

hair is short and she wears a diadem with a rosette at each end.

The mirror disk fits into the head of the statuette, and it is fastened by three rivets to a palmette-and-scrolls attachment at the back. The back of the mirror is decorated with an overall pattern of radiating curves, a popular decorative form for many mirrors. Along the edge is a tongue pattern in relief. The rampant griffin on either side gives additional support to the disk and rounds out the composition. The reflecting surface is slightly convex with a finely incised border. This mirror is 13 and ⅛ inches tall, and the diameter of its reflecting area is 5 ⅛ inches. Because the surface was corroded, Museum personnel made a polished metal disk of the same size and convexity. The reported that "instead of seeing barely your face, you could examine your hair, your necklace and the top of your dress. At a time when there were no large wall mirrors this was important."[3]

Although the mirror appears to be one that can stand alone, it cannot. The crouching lion on which the feet of the girl rest does not give sufficient support because the bottom is uneven. Mirrors of this kind were intended to be held, like the Egyptian mirrors from which they were derived.

Standing mirrors were a later development. These were introduced during the fifth century B.C. A most attractive style is known as the caryatid stand mirror. They are made of three parts: the reflecting disk; the draped, priestesslike figure, the caryatid composing the handle; and a base. Usually a fourth part, a support or transitional member uniting the disk and the caryatid, is present. A good example is shown in figure 12, which also shows the high fashion of the times. This mirror is 16 ⁷⁄₁₆ inches high. The female figure is dressed in a sleeveless Doric chiton, a garment commonly worn next to the skin, which falls in simple straight folds. Posing much as a modern high-fashion model might, she holds her right arm extended in front of her, with the palm of her hand outward. With the left hand she lifts a fold of her drapery. Her hair is arranged in simple fashion, parted in the

Greek caryatid mirror.

The Mirror and Man

front and coiled at the back around a small filet, which encircles her head. She supports the mirror through a decorative scrolled ornament with a rosette at each end and a lotus flower in the center. The back of the mirror is inscribed with concentric circles. The figure rests on a small round base supported on three lions' feet.

Mirrors were an indispensable article of the toilet for the Greeks, and every effort was made to keep them bright. When not is use, many were covered with cloth or straw. The more fashionable mirrors were provided with lids for that purpose. These lidded mirrors were most popular from the fifth century B.C. to about the third century B.C. Usually the cover is attached to the mirror by a hinge that permits it to be opened, like a modern-day compact. Sometimes a suspension ring is added to the hinge and another ring fixed to the opposite end for raising the lid. Many of the covers are decorated with a scene in relief surrounded by a molded border that matches the border of the mirror rim to give an appearance of continuity when the lid is closed. The inside of the lid is often ornamented with concentric circles, but sometimes the interior is polished like the mirror disk to serve also for reflection. Because of its concave shape, it was probably used to magnify portions of the face.

One of the most beautiful of all known Greek mirrors of that period is one of these compact mirrors. On the cover is a sculpturelike, high-relief portrait of a lovely woman. She has long, wavy hair which falls in loose curls about the head. Her brows and lashes and the shorter hairs on the forehead are delicately engraved. Her noble features suggest that she is one of the greater divinities, most likely Aphrodite, who appropriately serves as a decoration and an inspiration to the hopeful viewer. In her right hand she holds a lock of hair, a characteristic of depictions of Aphrodite.[4] This beautiful cover has a rim that fits on the beveled edge of the mirror disk, the diameter of which is about eight inches.

The elegant innovations of the ancient Greeks to the art of mirror design were modest compared with their monumental contributions to the science of mirrors. Their invention of geometry made it possible to establish a quantitative understanding of reflection that is the basis of modern mirror technology.

Scientific investigation began with Aristotle's examination of hearing and seeing. He noted first that the angle at which an echo rebounds from a hard surface, the reflected angle, is equal to the angle at which the sound strikes that surface, the incident angle. Aristotle probably reasoned from the analogous law of light reflection from a mirror. The latter is much more easily observed because of the distinctness and clarity of the incident and reflected rays. Support for this theory comes from Aristotle's observation that the phenomenon of echo and reflection are generally the same and from his comparison, in *Problemata*, of the reflection of light to a mechanical reflection of a bouncing ball.

The historian Carl Boyer notes that "[Aristotle's] works do not constitute explicit formulation of the optical law of reflection, but they do serve to make almost self-evident the inference that the equality of the angles, at least in the case of plane mirrors, was familiar to the Aristotelian school. . . . How long the law had been known can only be conjectured. . . . Plato . . . does not clearly establish such anterior knowledge. On the other hand, Aristotle touches upon the quantitative aspect of reflection in a manner so [casual] as to imply that in his day the discovery was by no means a recent one."[5]

The Greek engineer Heron of Alexandria provided an explicit formulation of the law of reflection about A.D. 100 in his book *Catoptrics*, the name applied to the branch of optics dealing with mirrors. Heron also documented for the first time a number of mirror illusions that were known and used by his knowledgeable predessors but never divulged. He himself devised a trick that was used by the priesthood to create a godlike apparition on the altar of a temple. With this, priests were able to control the mood of the unsuspecting populace and ensure loyalty to the ruling family.

In the 280 B.C. book *Catropica*, ascribed to Euclid, Proposition One states: "By mirrors whether plane, concave

The Mirror and Man

or convex, rays are reflected at equal angles," and the geometric diagram is given. This proposition seems to present the earliest publication of the law, but scholarly examination of the *Catropica* indicates that this matter was inserted later by Theon, the Alexandrian geometer, when he edited Euclid's original work in A.D. 400. Therefore the attribution of the proof of the law of reflection to Euclid is believed to be spurious.

Another important aspect of mirrors documented before Heron's time has to do with curved mirrors and their ability to gather light and concentrate it to a point. The precise shapes of these mirrors evolved fortuitously from the Greek geometers' study of conic sections. Apollonius made the first comprehensive study of conic sections. In a book attributed to him, *On the Burning Mirror*, he describes the focal properties of parabolic, or burning, mirrors.

A more specific source of information that sheds light on the Greek understanding of the mirror is the treatise *On Burning Mirrors*, a recently discovered work by Diocles from sometime between 196 and 180 B.C. Diocles claimed that no one before him had given a formal geometric proof of the focal property of the parabola, but he credited Diostheus, a mathematician of the third century B.C., as being the first to realize that the parabola had a focus. Diocles proved that a parabola has a focus and that a parabolic mirror concentrates the sun's rays to a point. He further proved that a spherical mirror concentrates the rays over a larger area and, therefore, was less efficient that a parabolic one because it ignited materials less readily.

Less than one hundred years after Heron, Ptolemy, the greatest astronomer of ancient times, wrote a five-volume treatise on optics. Two volumes deal with the theory of mirrors and contain experimental confirmation of the law of reflection. The last important Greek contribution of that era came from Anthemius of Tralles, an architect, geometer, and physicist. With a conic section he studied the mirror-reflection properties of the parabola and ellipse. Unfortunately, Greek science came to an abrupt end with the closing of the Athenian schools of learning in A.D. 529 by the emperor Justinian. Its legacy passed to the East, where Arabic scholars continued to contribute strongly to mathematics and optics.

A little-known chapter of early Grecian mirror history began in ancient Britain about the fifth century B.C. A few Greek mirrors were imported at that time, and the Celts began to develop bronze mirrors of their own, examples of which have been found in widely separated districts of Britain. These mirrors, made before A.D. 100, indicate the remarkable advances the Celts made in bronze metalwork. This skill had reached its peak before the Roman Conquest, 54 B.C., and in its best period was free of foreign influence. Celtic art is beautifully illustrated in the original curvilinear designs engraved on their bronze mirrors.

The first of these mirrors was found in a grave at Birdlip, Gloucestershire, in 1879 (fig. 13). The mirror is oval in shape measuring about eleven inches wide by ten inches long. The back is engraved with a intricate design in the form of a triple scroll with basketwork shading. Above the handle on both sides is a triple arrangement of trumpet-shaped scrolls in relief, with each of the three enclosed spaces decorated with two red enameled dots. The bronze handle is continuous with the border. It takes the form of a double loop ending in a ring that encloses a disk also decorated with two red enameled dots. Another mirror of almost the same description was found at Desborough and is now in the British Museum. Unfortunately, by about A.D. 100 the Celtic art of mirror design was on the decline. Fashion had changed, and the large native mirrors were superseded by the much smaller, less elaborate mirrors of their Roman conquerers.

Roman hand mirrors—the best were fabricated at Brundusium—were far less ornamental and elegant than those of Greece and Etruria, though they did follow the prevalent styles. Utility overtook aestheticism; engraved designs disappeared and concentric circles formed the only decoration. Mirrors were so popular they were owned even by servants. Mirrors and cosmetics application went hand in hand. Here the Romans appear to have surpassed the Egyptians, judging

The Mirror and Man

from Lucian's description of a Roman lady's toilet: "She must these days use powders, pomades, paints. . . . each chambermaid, each slave carries one of the essential objects for the toilet. One holds a silver basin . . . another . . . a water pot, still others the mirror."[6]

The Romans used a white bronze alloy richer in tin than the Etruscans or Greeks. They added a hard, silvery-white reflecting surface by hot-tinning the bronze surface. This consisted of laying a very thin sheet of tin over the face of the bronze mirror and heating the combination by fire to diffuse the tin directly into the underlying surface of bronze and form a bright reflecting surface.

The Romans were the first in Europe to produce glass mirrors, a technique they learned from their Egyptian subjects. In their first attempts, small, flat glass mirrors were backed with white plaster or resin, but the light reflected from these surfaces was dim and diffused. Later they tried a metallic backing for the glass that proved more successful. The metal-backed glass mirrors were convex. The glass was most likely blown as large bubbles ranging from ten to twenty-eight inches in diameter. After the bubble was blown, it was allowed to cool down to the melting point of lead. Then melted lead was poured in the concave side and swirled around until the inside of the bubble was set. Viewing was done from the convex side. At best these mirrors gave dark, distorted reflections that were poor substitutes for existing metal mirrors. Though little used, these glass mirrors were the primitive forerunners of the modern looking glass.

Nothing more was heard of glass mirrors in Europe until the twelfth century, except for a significant remark made by Isadore Bishop of Seville in the seventh century, recognizing the potential of glass in mirror making: "There is no material better adopted [than glass] for making mirrors."[7] In the larger sense, during the medieval period mention of the manufacture of mirrors or their use does not appear in the European literature of that time. There is an isolated reference to Pope Boniface IV sending a silver mir-

Celtic enameled bronze mirror.

ror and a comb of gilded ivory in A.D. 625 to the queen of Edwin of Northumbria.

It was not until the twelfth century in Alexander Neckam's *De naturis rerum* that there is a reference to mirror making. He alludes to mirrors of glass with a lead backing, commenting: "Take away the lead which is behind the glass and there will be no image of the one looking in."[8] He was probably aware of the glass mirror industries that had arisen in Germany and the Netherlands as a result of a German discovery that made flat, even, sheets of glass by a cylinder process. These flat sheets were coated with lead to make barely acceptable mirrors. The Venetians also began to produce glass mirrors. They were able to make clearer glass, but they had not yet learned the German method of sheet-glass production. German manufacturers also made small glass mirrors known as *Ochsenaugen* (Bulls eyes). These bear a striking resemblance to the early Roman convex mirrors. Globes of glass were blown, and while still hot, molten lead or tin was poured into the concave side. When the globe was entirely coated and cooled, it was cut into small convex pieces about one and one-half inches in diameter to make relatively good mirrors. That these glass mirrors were still inferior to metal mirrors is witnessed by the names used to describe their poor reflecting qualities. These were the Old Norse expression *Shattenche* (shadow seeing) and the Old High German *Shattengesicht* (shadow face), which aptly denote vague and dark images.

Metal mirrors were still popular in the days of chivalry and the troubadours. Their backs often had richly decorated scenes of love and gallantry, song and prowess, scenes that reflected the mind and spirit of those times. One of the most common representations was the siege of the Chateau d'Amour with attacking knights charging the fortress with roses. The gallant often carried a sword with a mirror ornately mounted in its hilt, which flashed in the sunlight as he rode to the defense of his lady.

Ancient and Medieval Europe: The Material Mirror

NOTES AND SOURCES

1. Eduard Gerhard and G. Karte, *Etruskische Spiegel*, 5 vols. (Berlin: 1840–97).
2. Nancy Thomson de Grummond, "Reflections on the Etruscan Mirror," *Archaeology*, Sept.–Oct. 1981, p. 57.
3. Gisela M. A. Richter, "An Archaic Greek Mirror," *Archaeological Institute of America* 42 (1938):341.
4. Gisela M. A. Richter, *Greek Etruscan and Roman Bronzes* (New York: Gilliss Press, 1915), p. 258.
5. Carl B. Boyer, "Aristotelian References to the Law of Reflection," *Isis* 36 (Jan. 1946):94, 95.
6. Richard Corson, *Fashions in Makeup* (New York: Universe Books, 1972), p. 60.
7. Bruno Schweig, "Mirrors," *Antiquity* 15 (1941):264.
8. Urban T. Holmes, Jr., *Daily Living in the Twelfth Century* (Madison: Univ. of Wisconsin Press, 1952), p. 144.

Other sources for this chapter include: J. D. Beazley, "The World of the Etruscan Mirror," *Journal of Hellenic Studies* 69 (1949):1–17; Morris R. Cohen and J. E. Drabkin, *A Source Book in Greek Science* (New York: McGraw-Hill, 1948); Lenore O. K. Congdon, "Metallic Analysis of Three Greek Caryatid Mirrors," *American Journal of Archaeology*, 71 (1967):149–53; S. H. Cuming, "On Mirrors," *Journal of the British Archaeological Association*, 1861, pp. 279–88; Mario A. Del Chiaro, "Etruscan Bronze Mirrors," *Archaeology*, Apr. 1974, pp. 120–26; Diocles, *On Burning Mirrors*, tr. G. J. Toomer (New York: Springer-Verlag, 1976); G. C. Dunning, "An Engraved Bronze Mirror from Nijmegen, Holland; with a note on the origin and distribution of the type," *Archaeological Journal* (London), 85 (Mar.–Dec. 1928):69–79; Winifred Lamb, *Ancient Greek and Roman Bronzes* (Chicago: Argonaut, 1969); E. A. Lane, "An Etruscan Bronze Mirror in the Victoria and Albert Museum," *Journal of Hellenistic Studies* 57, pt. 2 (1937):219–23; Carlo and Leoni Massimo Panseri, "The Manufacturing Technique of Etruscan Mirrors," *Studies in Conservation*, 3 (1957):49–62; Ernest A. Parkyn, *An Introduction to the Study of Prehistoric Art* (London: Longmans, Green, 1915); Agnes C. Vaughan, *Those Mysterious Etruscans* (London: Hale, 1966).

ANCIENT AND MEDIEVAL EUROPE
The Figurative Mirror

A s Christianity spread throughout Western Europe during its first thirteen centuries, it made powerful use of the mirror to disseminate its teachings. Philosophers and theologians used the mirror as a figure or metaphor to strengthen the concept of God and ideality, purity and wisdom, morality and true self-knowledge. There also developed a mirror dualism which Dante alluded to in his *Divine Comedy*, where an "inferior" mirror obeyed the natural law of reflection and a "superior" mirror received the direct illumination of God.[1]

The origin of the mirror as a figure in early Christianity begins with a tale of a man and his mirror told by Seneca. The Roman Empire then, about A.D. 40, was rich and powerful. Luxurious living in the estates of the affluent and mighty was enhanced by full-length mirrors carved of gold and silver and adorned with jewels. They reflected the personal vanity of a jaded general or a pompous prince. The mirror, which had been an ornament of women, now became a desired accessory of the soldier. For others, the mirror was a device to reflect debauchery.

Seneca grumbled that the mirror was no longer an article for good grooming but had become an implement of vanity and vice. He told of a rich and greedy man named Hostius Quadra, who satisfied his inordinate lust by the use of mirrors: "He had mirrors made of the type [concave mirrors] in which a finger exceeded the size and thickness of an arm. These, moreover, he so arranged that when he was offering himself to a man he might see in a mirror all the movements of his stallion behind him and then take delight in the false size of his partner's very member just as though it were really so big. . . . Mirrors faced him on all sides in order that he might be a spectator of his own shame."[2] Later in the narrative Hostius Quadra says, "I will surround myself with mirrors, the type of which renders the size of objects incredible. If it were possible, I would make those sizes real; because it is not possible, I will feast myself on the illusion. Let my lust see more than it consumes and marvel at what it undergoes."[3] Hostius' lust was so immense that nature itself could not satisfy it, and he made mirrors the unnatural means of deceiving himself into an illusory satisfaction.

Seneca condemned all this and reminded his readers that mirrors were invented so that man might know himself, gain knowledge of himself and thereby wisdom. Thus through the mirror the handsome man may avoid infamy; the young man be reminded that youth is a time of learning, and "the old man, set aside actions dishonourable to his gray hair, to think some thoughts about death. This is why nature has given us the opportunity of seeing ourselves."[4]

Elsewhere Seneca comments on nature's purpose in creating real objects and wanting reflections of them to be seen. "Surely," he says, "it is not in order that we men may pluck out our whiskers in front of a mirror or make our faces smooth."[5] He believed that the mirror was made for a grand purpose: to allow man to see the sun's form with its brightness dimmed by viewing it by reflection in a mirror. (This indicates the poor reflecting qualities of mirrors of that time.) Man, who would be blinded by looking into the sun directly, would be ignorant of nature's central power; and nature, wanting men to see this wondrous creation at least indirectly,

The Mirror and Man

has invented reflection either by a liquid or metal to make this possible.

Underlying this thought was the idea that reflection in a mirror was only a poor imitation of the real thing. It echoes the profound insight of Plato in the *Republic*. About 375 B.C. Plato wrote of the distinction between true forms and images, ultimate truth and illusion. Illusions, he believed, were impressions and opinions of which the minds of ordinary people are full, no more than a shadow or reflection of the ultimate truth or ideal. Plato's philosophy extolled the pure idea created by man as the highest accomplishment of which all physical existences are only imperfect copies. For example, he regarded moral beauty as the prototype of which all visible beauty is only an image. In speaking of the inferior copy or image of the ideal, he considered the mirror, with its poor reflection, the perfect device to objectify this concept. Initiating the analogy later used by Seneca, Plato argued that it was often impossible to see the ideal, just as the true nature of the sun, the source of all life, cannot be viewed directly because of the blinding effect it has on the observer. He can only see it dimly and imperfectly by reflection in a mirror. He extended his philosophy by noting that if there was a material world known through our senses, then it was only a reflection of the "other world" of our ideals which we cannot see, feel, or hear unless possessed of universal wisdom.

An echo of this Platonic mirror occurs in the seventh chapter of the second century B.C. *Book of Wisdom* where wisdom is defined:

> For she [wisdom] is a reflection of eternal
>   light
> And a spotless mirror of the working of
>   God,
> And an image of his goodness.[6]

The Christian view of this concept is the Pauline mirror.

To the church in Corinth, a center of mirror production, Paul in A.D. 54 addressed his Epistle in which he remarked,

"For now we see in a mirror dimly, but then face to face. Now I know in part; then I shall understand fully, even as I have fully understood." The opening is more familiar in the Authorized (1611) translation: "For now we see through a glass, darkly." (1 Corinthians 13:12).

Paul was no doubt aware of Plato's idea that reality is only a poor reflection of the ideal and adopted his mirror analogy, replacing the sun as the life source with the God figure. Using this theme, Paul expressed the imperfect nature of human knowledge in this life. Man may see the glory of God only as in a mirror, that is, dimly, never directly. Man must await his translation to, and acceptance by, the spiritual realm before he can feast his astonished eyes upon the true nature of God in heaven, which he now sees as an imperfect image.

From these beginnings the mirror analogy took on great significance in Christian thought. From the phrase "spotless mirror" (*speculum sine macula*) in the *Book of Wisdom* was drawn the attribute of the purity of the Virgin Mary. This concept, which associates the Virgin's purity with a spotless mirror, lasted through the sixteenth century. It was strengthened in the thirteenth century by Jacobus de Varagine. He wrote that the Virgin was called a mirror because of the mirror's composition of glass and lead. Glass represented her virginity because, as the sun penetrates glass without violating it, so Mary became a mother without losing her virginity. Lead symbolized her ductility, and the ashen color of the mirror signified her humility. "She is called a mirror because of her representation of things, for as all things are reflected from a mirror, so in the blessed Virgin, as in the mirror of God, ought all to see their impurities and spots, and purify and correct them; for the proud, beholding her humility, see their blemishes, the avaricious see theirs in her poverty, the lovers of pleasures, theirs in her virginity."[7]

The Platonic philosophical concept of the mirror prevailed for over fifteen-hundred years in Europe, not only in religion but in art and morality.

About two-hundred years after St. Paul, the Platonic

The Mirror and Man

view that the world is a reflection of the ideas in the mind of God was revived by the Neoplatonists and transmitted successfully to the Middle Ages because it confirmed current Christian doctrine. The metaphor of the mirror lent itself to the Christian belief that all existence is understood as a relation between paragon and image, between a reality and its innumerable reflections. Contempt for the world of matter and belief in the liberation of the soul through asceticism and mystic revelation made this especially congenial to many believers. Plotinus noted that:

> as in a mirror, the semblance is in one place, the substance in another, so Matter seems to be full when it is empty, and contains nothing while seeming to contain all things. The copies and shadows of real things which pass in and out of it, come into it as in a formless shadow. They are seen in it because it has no form of its own.
>
> In its passivity and formlessness it can produce nothing but an illusion. The mirror of matter represents the delusiveness of experience in the sensible world, a world of shifting images, a world whose substratum has the "actuality of an illusion."[8]

The Neoplatonists used another mirror analogy known as the chain of being to explain the process of creation as emanating from God down to the lowest creatures on earth. They had found a concept expressed in the *Timaeus* that dealt with the generation of the Many from the One. Their idea was that God's radiant image was reflected sequentially in a chain of mirrors; the first reflection was the creation of Mind, the second was the creation of Soul, and each subsequent reflection created life from its highest to lowest forms according to the degeneration of God's reflection until the lowest mirror reflection, which represented the last dregs of things. Man represented a fairly good image of God whereas other creatures, depending on their place in the hierarchy, were successively poorer reflections. Theologians accepted this idea over the next millennium.

Ancient and Medieval Europe: The Figurative Mirror

During the fourth century the mirror analogy became firmly embedded in Christian theology through the writings of St. Basil of Caesarea, St. Gregory of Nyssa, and most powerfully through the work of St. Augustine. Basil and Gregory likened the purity of the soul to a property of the mirror in somewhat different ways, but their messages were clear and they point in the same direction. Basil believed that the pure soul is the mirror of God. Keep this mirror clean he warns, for "from a soiled mirror you cannot get images; nor can the soul that is filled with worldly cares and over which the flesh spreads darkness receive the illumination of the Holy Spirit."[9] Gregory, writing of the purity of the soul, describes it as a mirror able to be turned to reflect either the wayward world or the superior world: "When one's soul, after the manner of a mirror, has turned from what has seduced him to evil toward the hope of future good, he can see in the purity of his own soul the forms and images of beauty shown to him by Divine aid."[10] The mirror here is ambivalent.

Augustine used the mirror metaphor on a much grander scale than his predecessors, yet it represented only one element of his profound metaphysical conceptions, which effectively synthesized Christian theology. His word was gospel, for it was accepted by Western scholars well into the Renaissance. Augustine repeatedly praised the Platonists above all other philosophers, and he was indebted to them for many of his metaphysical ideas, including his mirror analogies. Manichaeism, the "Religion of Light" to which Augustine adhered in his youth, also influenced him with its equating light and darkness with good and evil and an associated symbolism.

Augustine, echoing Paul, argued that the obscurity of the mirror reflected the sacrosanctity of God. This experience, he felt, is valuable in itself, for it purifies the mirror that reflects God's image. When the vision "face to face" that is promised the true believer comes, he will see God with greater clarity and sureness than he now sees his image. Conversely, "those who see the trinity in their mind but do not believe or understand it to be an image of God, those people indeed see a mirror, but they fail to see through the mirror the God who must be seen there; they do not even know that the mirror they see is a mirror. If they did they would seek through that mirror for the One whom it reflects, in order that, their hearts cleansed by a faith unfeigned, they might see him face to face."[11]

In his commentary on Psalm 103 Augustine refers to Scripture as a mirror, giving it a twofold function: in its resplendence it shows you what you will be, that is, pure of heart, and it also shows you what you are, that you may confess your deformity and begin to adorn yourself.[12]

The ambivalent nature of the mirror appears again when Augustine uses its properties to demonstrate that "the matter of falsity is the similitude of things which reaches the eyes." The image in the mirror is a "seductive similitude" seeming to be real, but it is really false; for it cannot be grasped, it makes no sound, it does not live. "For does not your image in the mirror appear to you as though it wished to be yourself, but to be false precisely because it is not?" He further states that within us is the desire to see the whole face of truth, but often we are deluded, thinking that what we know or seek to know is all that is to be known. From our partial knowledge "certain false colors and forms pour themselves as though into a mirror of thought. . . . Such Imaginations are to be avoided with great care; they are recognized as false when they vary as if in a varying mirror of thought whereas the face of truth remains one and immutable."[13] The mirror is the natural example of instability, passivity, and delusiveness.

Through the writings of Augustine and, to a lesser extent, those of St. Basil and St. Gregory, basic theological mirror analogies were formulated: the mirror of the soul and the mirror of the mind. The mirror of the soul was the image of the ideal, or archetypal idea. This led to mirrors known as compendiums of knowledge and idealized virtures. The mirror of the mind was ambivalent; it reflected the shadow world of the senses and led to mirrors that warned of the transience and illusiveness of this world. On the other hand, it concerned itself with religious truth. In this sense Holy Scripture was called a mirror from which could be drawn models of holy living either as biography or as religious rule. The common mandate in these was to judge changing appearances in the light of basic ideas, to seek after ideal beauty and unchanging truth; that is, to seek after wisdom.

Serious philosophical speculation about these Augustinian ideas began in the twelfth century when the idea that "the true mirror is the highest Wisdom" was amplified in great detail by Alanus ab Insulis. His metaphysical reasoning contained a rationale for the ambivalent mirror. He constructed a hierarchy of truth as found in the Scriptures, in human nature, and in creatures. These he represented as a threefold mirror, one in which you ought to look to see what you ought to be. "Hence, in the mirror of the Scriptures you see your present state; in the mirror of the creatures, you see yourself as the wretched one; and in the mirror of your human nature you judge yourself as guilty."[14]

From your nature, he continued, there arises another threefold mirror, one of reason, sensuality, and the flesh. Of these, reason is the true mirror wherein you see things correctly. In the false mirror of sensuality these things are reversed, and in the most false mirror of the flesh, they appear completely inverted. All three mirrors were to be studied: reason that you may obey it; sensuality that you may subject it to reason; and carnality, that you may chastise it.

For the subjecting of the flesh, three more mirrors were proposed: of providence, that you may regard what is before you; of circumspection, that you may be alert to what is opposed to virtue; and an unlabeled third mirror, where you will learn how to avoid the hypocrisy of cruelty, of feigning justice, and of indulgence masquerading as mercy. This final threefold mirror will bring you three visions: a pure conscience, divine contemplation, and eternal life.

About one hundred years later, St. Bonaventura devised a dualism with the image of a mirror to distinguish between earthliness and divinity. He wrote of the mirror of creation (*speculum inferius*) and the mirror of God (*speculum superius*), dealing for the most part with the former concept. For him

The Mirror and Man

the mirror of creation produced a hierarchy of reflections from God downward. This was the chain of being. But his conception of it was more imaginative: "An angel is also a pure and most clear mirror; although it is spotless in its act, it is nevertheless sullied in its potency, being at least at some remove [from the level of God's power]. The human spirit is also a mirror, sullied in its act and potency, in which the divine image shines. All creation is said to be a mirror, also, since the invisible things of God are represented in it, as it were through a trace of Him"[5]

For St. Bonaventura, the angels, pure and unpolluted by matter and change, afford the clearest reflections of God. The clarity of the angelic mirror is flawed in only one respect. It does not quite equal God's power. Only God has the supreme power to create from nothing, to bestow existence and the forms conceived in the Divine Mind. Angelic mirrors can reflect but never create.

Compared with the angels, man is a far inferior mirror: he is tarnished because his soul is restricted to his body, which is subject to dissolution. Despite these imperfections, man still manages to reflect the divine image because God directly illuminates the intellectual faculty of the human soul and allows it a dim perception of unchangeable truths.

The lowest mirrors are material things that possess no rational soul. Consequently they cannot produce an image of God but only a faint trace. Like Paul, Bonaventura regarded the universe as a "glass" through which we see God "darkly."

Compared with the mirror of creation, the mirror of God was unique, and Thomas Aquinas, a contemporary of Bonaventura, considered two ways of looking at it. First, we might sense the divine mirror as an object in itself provided that we have reached the ultimate perfection granted only to the sanctified when the essence of God is fully revealed. A second way is to observe the universal forms of created things reflected in God's mind. Of course, the mirror of God does not reflect passively, for this would imply that God receives light from a higher authority. Instead, God's mind reflects the universal forms of things in an active sense, as if a mirror were suddenly able to create images from its inherent radiation. Images produced by the divine mirror are more properly called exemplars than reflections. This metaphor helped suggest the similarity between particular created things and their universal forms. According to Thomas, the superior mirror represents God's foreknowledge, which the blessed could gaze upon in order to perceive the essence of things or the course of future events.[16]

During this period the interpretation of the *speculum superius* was acceptable because of its association with the metaphysics of light. Light was defined by Augustine as the divine illumination of God. Proponents of this doctrine, such as Robert Grosseteste, established a plausible distinction between two kinds of light: *lux* and *lumen*. *Lumen* was reflected light visible as a natural phenomenon. *Lux* was primal light radiating directly from God. It possessed the power to create forms as the Divine Mind deemed. Although distinguishable, these two kinds of light were similar to each other and not in opposition. The correspondence between *lumen* in the real world and *lux* in the spiritual world allowed the metaphysicians of light the comparison of God to a mirror by defining reflected light as coming from a material mirror and primal light emanating from a spiritual mirror.[17]

By the close of the thirteenth century a paradox in the understanding of the figurative mirror had developed. The material nature of the mirror contradicted its spiritual essence, or ideal reality. It depended on whether man viewed the mirror with his human frailties or transcended his mortal failings to view the mirror with his soul.

Reflection seemed to possess a dual personality. The act of looking into the mirror, and the mirror itself as an object, gathered opposing symbolic associations. The mirror could represent purity, truth, and wisdom. The unblemished mirror, *speculum sine macula*, was a symbol of the purity of the Virgin and of Christ and of chastity. It became popular with the great revival of worship of the Virgin that spread over the Christian world in the twelfth century. This iconography appeared as late as the sixteenth century on one of the seventeen tapestries of the Rheims cathedral, representing the perfections of the Virgin according to the litany. In other iconography, Prudence or Wisdom, because she sees so clearly, has a mirror as one of her attributes, and her looking into it at her own face suggests the search for self-knowledge undertaken by the wise. On the other hand, excessive mirror gazing could indicate an obsession with one's self, and so the mirror was also associated with the two deadly sins: Superbia, or Pride, and Luxuria, or Lust. Both exemplified the characteristics of the evil mind. Medieval artists frequently portrayed these personifications with mirrors in their paintings or in stained glass windows in the great French cathedrals, such as Notre-Dame. This symbolism continued to be used by artists into the Renaissance, but it withered away by the sixteenth century. Hieronymus Bosch's *Superbia* (fig. 14) reenforces an evil connotation by having the devil hold up the mirror to the conceited woman. He mimics her grotesquely by wearing the same headdress. Pieter Brueghel's *Superbia* shows an elaborately dressed woman looking conceitedly into a mirror attended by her peacock, the emblem of the deadly sin of Pride (fig. 15).

In Laux Furtenagel's well known portrait of Hans Burgkmaier and his wife, painted in 1527 (fig. 16), the mirror is simultaneously associated with *Vanitas*, *Veritas*, and *Prudentia*. Instead of the couple's faces, two skulls appear in the mirror. The inscription in the upper right-hand corner emphasizes the strange apparition: "This is what we looked like—in the mirror, however, nothing appeared but that." The inscription on the mirror's frame, "Know Thyself" points to Self-Knowledge which, together with Retrospection and Foresight, are the implications of the mirror in the hands of Prudence.

An interesting extension of these metaphors developed from the mirror's association with the Holy Scriptures as referred to in the writings of Alanus ab Insulis. Holy Scripture was a mirror of what one ought to be. This mirror exemplified the word of God and the pattern of perfection.

A good example is found in the twelfth-century Latin

Hieronymus Bosch, Superbia *(from the painted tableboard with the Seven Deadly Sins). (Museo del Prado)*

Cock excud cum priuileg 1558

NEMO   SVPERBVS   AMAT

Houerdije   werdt   van   godt   bouen   al   gh

SVPERBIA ·

ROS · NEC AMATVR AB ILLIS ·

After Pieter Brueghel, Superbia (engraving). (The
Metropolitan Museum of Art, Harris Brisbane Dick Fund,
1926. All rights reserved, The Metropolitan Museum
of Art)

Laux Furtenagel, Hans Burgkmaier and His Wife.
(Kunsthistorisches Museum, Vienna)

manuscript entitled *Speculum Virginium* by an unknown writer under the pseudonym of Peregrinus. Peregrinus explains the term *speculum:* maidens look into mirrors, he says, to see whether there is any increase or decrease of their adornment, but Scripture is a mirror from which they may learn how they can please the eternal spouse. In this mirror they can find themselves and understand what they ought to do and what to avoid. Twelve chapters in his book give instruction on this matter.[18]

From this concept the term *speculum*, or mirror, was applied to a book, which by showing the truth of things, could act as guide and admonition to its readers. Such "mirrors," or "specula" were instructional and were used to establish models of behavior to be followed or avoided. Initially, they dealt with theological examples for the pious to follow, but as the speculum genre came into universal use from the twelfth to the sixteenth century, religious themes were supplanted by profane ones dealing with topics ranging from governing, such as *A Mirror for Magistrates*, to etiquette, as *The Mirour of Good Manners.*

A Mirror for Magistrates (1559) provides a guide for the ruling class. The dedication points out that the success or failure of any realm lies in the virtues or vices of its rulers. Peace and prosperity, disorder and misery, come through the just or unjust actions of rulers. "For here in a loking glass you shall see (if any vice be in you) howe the like hath bene punished in other heretofore, whereby admonished, I trust it will be a good occasion to move you to the [quick correction of faults]. This is the chiefest ende, whye it is set furth, which God graunt it may attayne."[19]

In the thirteenth century a "mirror" of monumental literary significance challenged existing mores by dealing with the delicate subject of love in a forthright fashion. *The Roman de la Rose* was a poem in two parts of over twenty-two thousand lines begun (and left unfinished) by Guillaume de Lorris (4,058 lines, c. 1237) and continued by Jean de Meun (c. 1275–80). The poem could also be called a 'Mirror for Lovers" because its aim was to set forth the "whole Art of

Love." Part One is an often subtle elaboration of the psychology of love, in the form of the then fashionable allegorical dream vision. In Part Two love remains an important subject, but the conventions of courtly love are replaced by realistic views, and the poem is expanded to comprehend virtually the thought of the age.

Jean de Meun, sometimes called the "Voltaire of the thirteenth century," had mastered all the scientific and literary knowledge of his day in France. His part of the *Roman* deals with a theological mirror metaphor, but he also alludes to the physical properties of the mirror to develop this subject. Nature gives a discourse on optics to indicate the powers of the mirror known in the thirteenth century. Then the author leads his reader into metaphysics to construct a mirror through which one might look and pierce the smoke screen that so often is thrown about the simple realities of human passion. His mirror would not be like those that shroud the truth by creating delusive images.

Jean de Meun's rational treatment of love enables the reader to witness the consummation of the sexual act. This mirror fulfills the promise to make unmistakably clear that the *Roman* is about "leaving nothing of the thing hidden."[20] Yet, like all human mirrors, the "Mirror of Love" reflects only images that are transitory. So the images that lovers see when they look into the speculum of this poem are images of their own transitory and corruptible nature. The lovers to whom the poem is addressed may use the mirror to inflame their passions or inform their reason.

As part of the act of love Jean de Meun used the mirror as a symbol of human reproduction, tying it analogously to the chain of being. The chain of being can be considered as a chain of mirrors dependent from God, the successive members of which reflect God's image with less and less distinctness. On a lesser scale, the reflection of an image within a mirror is a distinct analogy to reproduction. As a son is called the "image of his father," so all God's creatures (who are his children) may be called his images. The mirror, therefore, which symbolizes the radi-

Ancient and Medieval Europe: The Figurative Mirror

ant source of life also symbolizes the process of generation of the individual.

Jean de Meun's ulterior aim is to encourage man to generate true and living images of God. To emphasize his purpose, he chastises those who generate false and lying images, those who live in a world of hypocrisy, superstition, and idle fancy. He asks the reader to enter the world of true and faithful images, each of which is reflected from the everluminous mirror which is the Mind of God.

When Jean de Meun used the natural properties of the mirror as the first step toward arriving at his metaphysical one, he brought into focus the conflict of the thirteenth century. During this period there was an awakening of the experimental method. It was launched by Roger Bacon to test theories relating to natural phenomena and to question theological beliefs. This growing dilemma was handled by the kind of dualism expressed in Robert Grosseteste's *lumen* and *lux* theory and St. Bonaventura's conception of the natural mirror and the divine mirror.

Roger Bacon found a way to overcome this by skillfully justifying the study of the physical mirror through metaphysical reasoning. To appreciate fully Bacon's efforts, it is useful to review his background and his scientific knowledge. He benefited from the revival of Greek and Arabic scientific contributions. These contributions reached their zenith between A.D. 900 and 1100, and they revolved about an Arab mathematician.

Ibn al-Haytham was born in Basra, Mesopotamia, in 965 and died in Cairo in 1039. Known as Alhazen in the West, he is considered one of the greatest contributors to the quantitative understanding of optical phenomena. The main body of his optical knowledge is represented by the *Kitab al-Manazer*, which in 1572 was first printed in Europe in Latin under the title *Opticae Thesaurus*. It had been known in Europe since the twelfth century.

Alhazen's work was the prime Arabic contribution to the optics of mirrors, and it laid the foundation for the early European efforts that followed. The first European to explore this field was the Polish scientist Witelo. He wrote elaborately on the focusing properties of the parabolic mirror and the technique for its manufacture. Indeed, optics in Western Europe from the thirteenth through the sixteenth century was dominated by the works of Alhazen and Witelo, who wrote *Perspective*.

Bacon learned from these works. Moreover, he revolutionized scientific thought by formulating the concept that science can be pursued only by direct observation of nature, that is, inductively, and by holding that authority in science was a stumbling block. In this he was extending the thought of his predecessor Robert Grosseteste, who stressed the importance of experimental verification in science. They both used optics, including mirrors, as the science in which they developed their ideas. Despite their intellectual approach, they still preserved a dualistic outlook regarding the nature of things.

Bacon's dual outlook may have been politically inspired. When Pope Clement IV asked him in 1260 to write on the scientific notions of his time, his first problem was the attitude he should take to new knowledge. He knew full well that the church's view toward science was authoritarian. To show that natural science, far from being a danger to Christendom, was a source of both wisdom and power, Bacon called up the ideas of Augustine, who had made ample use of mirror metaphors in his theological discourses. Bacon adopted from Augustine the view of seeing in light, and the nature of its reflection, the analogy of divine grace and the illumination of the human intellect by divine truth. Thus, he reasoned, the study of light was the godly key to nature and truth. With this, Bacon justified his support for the scientific approach in his *Opus Majus*, which he published in 1268. This was truly the major scientific work of that time. His section on optics was studied at Oxford at the end of the thirteenth century, and it was used there for over 300 years.

Bacon also discussed the concave spherical mirror, a favorite of magicians from earliest times. He wrote: "Among all mirrors the greatest deception is in the concave spherical ones . . . deception in regard to size . . . deception in regard to number. . . . Moreover . . . the object appears sometimes erect, sometimes inverted; and thus it appears that in these mirrors nothing appears except with deception."[21]

Bacon describes how an enemy may be outwitted by mirror-produced illusions of figures appearing in the air so that when one approaches to do battle with this army there are only images. He concludes that "in accordance with such principles as have now been touched upon in regard to reflection . . . not only could results be attained advantageous to our friends and terrible to our enemies, but very great comforts can come to us from philosophy, so that every deceit of jesters may be dimmed by the beauty of the wonders of science and men may rejoice in the truth, banishing far from them the tricks of the magicians."[22]

With this peroration Bacon deftly leads his readers, undoubtedly with Pope Clement in mind, into accepting the virtues of science by noting its ability to protect the state, establish truth, and overcome falsity. Through reflection the seeds of rationalism had been planted to flower through the Renaissance and thereafter.

In the meantime, Dante, the great poet of the period, very much aware of the natural mirror and the divine mirror, wrestled with these in the *Divine Comedy*. There he emphasizes the divine awakening as his Soul travels from hell through purgatory to heaven. When in purgatory, Dante's soul, guided by Virgil, sees light reflected from a mirror surface and explains it in the experimental, human sense that the angle of incidence equals the angle of reflection:

> *When a ray strikes a glass or water, its reflection*
> *leaps upward from the surface once again*
> *at the same angle but opposite direction*
> *from which it strikes, and in an equal space*
> *spreads equally from a plumb-line to mid-point*
> *as trial and theory show to be the case.*[23]

As one not exposed to heaven, he sees only the scientific, or worldly, result. When he finally enters paradise with

The Mirror and Man

Beatrice as his guide, a more enlightening experience befalls him when he asks Beatrice to explain the moon's appearance. Presumably she explains this with an experimental arrangement of three mirrors and a hidden light source:

> *Take three clear mirrors. Let two be set out*
> *at an equal distance from you, and a third*
> *between them, but further back. Now turn about*
> *to face them, and let someone set a light*
> *behind your back so that it strikes all three*
> *and is reflected from them to your sight.*
>
> *Although the image from the greater distance*
> *is smaller than the others, you must note*
> *that all three shine back with equal brillance.*[24]

James L. Miller has analyzed this event in his article "The Three Mirrors of Dante's Paradiso." Beatrice, far from praising the experimental method as exemplified by the equal brightness of the remote mirror, belittles it as the "only source" of Dante's knowledge, emphasizing the contrast between human knowledge and divine revelation. Man's knowlege, dependent on mundane experimentation, is incomplete. Beatrice's knowledge is a result of her heavenly experience reserved for the soul directly illuminated by God.

The structure of the *Paradiso* is framed by imagery involving mirrors and two "experiences": a physical experiment for its humble prelude and a spiritual experience for its exalted finale. Dante compares God to a "truthful mirror" incapable of distorting images. The force of his metaphor is intensified by contrast with the typical mirrors of his day.

The third mirror in Beatrice's description stresses the unvaried brightness of reflected light, for the fire therein shines as brightly as in the first and second. Dante, who initially looks at the moon's appearance as a problem in optics, discards his assumptions when instructed by Beatrice, and he discovers the divine truth regarding the different capacities of God's creatures to receive his illumination.

Ancient and Medieval Europe: The Figurative Mirror

As Miller points out: "Generalizing from experimental observations, [Beatrice] reasons that the diverse potencies in created things emanate from a single source of light diffused from the Prime Intelligence to the stars and below. Interpreted allegorically, the concealed light of the experiment corresponds to the divine light illuminating the universe; the mirrors represent the multiplicity of created things, receiving the divine illuminations and reflecting the light of the Creator. Creation becomes a hierarchy of mirrors, each casting an image of God. The further off the individual mirror is from God, the smaller the image of Him it reflects, but his brightness never diminishes."[25]

Dante alludes throughout the *Paradiso* to the mirror of creation (*speculum inferius*) and the mirror of God (*speculum superius*). Dante's pilgrimage through heaven to the throne of God may be seen as a transition from the inferior to the superior mirror. "Having transcended the speculum inferius, Dante is now prepared to receive the direct illumination of God. No longer must he see 'through a glass darkly.' "[26]

The bond between theology and the mirror was beginning to weaken, but not enough for philosophers and the writers of the time to reject more than a thousand years of teaching. But the great improvements in the art of mirror making over the next three hundred years would finally separate the mirror from these theological associations.

NOTES AND SOURCES

1. James L. Miller, "The Mirrors of Dante's Paradiso," *University of Toronto Quarterly* 46 (1977):266.

2. Seneca, *Naturales Quaestiones*, tr. Thomas H. Corcoran (Cambridge: Harvard Univ. Press, 1971), 1. 16.2–3.

3. Ibid., 1.16.8–16.9.

4. Ibid., 1.17.4.

5. Ibid., 1.17.2.

5. *Book of Wisdom*, tr. Joseph Reider (New York: Harper, 1957), p. 117.

7. E. C. Richardson, *Materials for a Life of Jacopo da Voragine*, pt. 2 (New York: H. W. Wilson, 1935), p. 66.

8. Frederick Goldin, *The Mirror of Narcissus in the Courtly Love Lyric* (Ithaca: Cornell Univ. Press, 1967), p. 6.

9. Sister Ritamary Bradley, "Backgrounds of the Title *Speculum* in Medieval Literature," *Speculum* 29 (1954):108.

10. Ibid., p. 107.

11. St. Augustine, *The Trinity* (Washington: Catholic Univ. of America Press, 1963), bk. 11, chap. 24.

12. Bradley, "Backgrounds," p. 107.

13. Goldin, *Mirror of Narcissus*, p. 7.

14. Bradley, "Backgrounds," p. 112.

15. Miller, "Three Mirrors of Dante's Paradiso," p. 267.

16. Ibid., p. 272.

17. Ibid., p. 271.

18. Arthur Watson, "The Speculum Virginium with Special Reference to the Tree of Jesse," *Speculum* 3 (1928):446.

19. *A Mirror For Magistrates*, ed. Lily B. Campbell (Cambridge: Cambridge Univ. Press, 1938), p. 65.

20. Patricia J. Eberle, "The Optical Design of the Romance of the Rose," *University of Toronto Quarterly* 46 (1977):259.

21. Roger Bacon, *Opus Majus*, 2 vols., tr. Robert B. Burke (New York: Russel and Russel, 1962), 1:553.

22. Ibid., p. 582.

23. Dante, *The Divine Comedy*, tr. John Ciardi (New York: Norton, 1977), *Purgatorio*, XV, 16–21.

24. Ibid., *Paradiso*, II, 97–105.

25. Miller, "Three Mirrors of Dante's Paradiso," p. 266.

26. Ibid., p. 273.

Other sources for this chapter include: Leroy Appleton, *Symbolism in Liturgical Art* (New York: Scribners, 1959); George A. Buttrick et al., eds, *The Interpreters Bible*, vol. 10 (New York: Abingdon-Cokesbury, 1953); Guillaume de Lorris and Jean de Meun, *The Romance of the Rose*, tr. Charles Dahlberg (Princeton: Princeton Univ. Press, 1971); Alan M. F. Gunn, *The Mirror of Love: A Reinterpretation of "The Romance of the Rose"* (Lubbock: Texas Tech. Press, 1952); Plato, *The Republic*, tr. Desmond Lee (Harmondsworth: Penguin Books, 1974); Archibald Robertson, *International Critical Commentary*, vol. 33, 1 Corinthians 13:12, (New York: Scribners, 1911); James Williams, "Mirror and Speculum in Book Titles," *Law Magazine and Review* (London) 26 (1901):157–63.

8

THE RENAISSANCE

T HE RENAISSANCE IS conventionally delimited as starting
early in the fourteenth century and ending at the close
of the sixteenth century. From the standpoint of the development of the mirror and its influence on art and literature, the
period extends to the end of the seventeenth century. The
modern glass mirror, or looking glass, was born in the
Renaissance. The metal mirror of the past virtually disappeared. During this period the Venetians discovered the process for making clear, colorless glass and the means of backing it with a bright silvery reflecting surface to produce brilliant images never seen before. Good, flat glass mirrors were
made that produced clear and accurate reflections. This new
clarity and fidelity of the mirror reflected the spirit of the
Renaissance, which saw a philosophy of world reality and
natural clarity overtake the metaphysical world of religion
seen "through a glass, darkly." The virtues of this new looking glass were quickly adopted by painters and led to an
altered metaphoric application by the poets of the time, but
the moral implications of self-admiration still clung to the
mirror throughout the period.

We now believe that glass was discovered accidentally in
ancient Syria about 2500 B.C. Pliny was the first to set down
an account of this discovery in his *Natural History:*

> That part of Syria which is known as Phoenicia
> and borders on Judea contains a swamp called Candebia

on the lower slopes of Mt. Carmel. This is believed to be the source of the River Belus, which, after traversing a distance of five miles, flows into the sea near the colony of Ptolemais (Akko). . . .

The river is muddy and flows in a deep channel, revealing its sands only when the tide ebbs. For it is not until they have been tossed by the waves and cleansed of impurities that they glisten. . . . The beach stretches for not more than half a mile, and yet for many centuries the production of glass depended on this area alone.[1]

There is a story that once a ship belonging to some traders in natural soda (natron, an Egyptian product) put in here and that they scattered along the shore to prepare a meal. Since, however, no stones suitable for supporting their cauldrons were forthcoming, they rested them on lumps of soda from their cargo. When these became heated and were completely mingled with the sand on the beach a strange translucent liquid flowed forth in streams; and this, it is said was the origin of glass.[2]

Pliny obtained most of his information from Greek accounts. The Greeks learned much about Phoenicia when they began trading there during the seventh century B.C. Greek merchants sought articles of glass from the Phoenicians in Sidon, and this port became synonymous with glassmaking. During many years of trading, the Greeks brought back stories about the accidental discovery of glass from the banks of the fabulous river Belus, now called the Na'aman, with its magical sand.

Pliny does not say that the Phoenicians discovered glass. He refers only to some traders in natural soda, natron, and there lies the unsuspected yet critical connection with Egypt as the supplier of the key ingredient for glassmaking upon which the modern mirror depends.

Natron was also used for mummification, the well-known ancient Egyptian practice that spread to its colonies. One Egyptian colony in Phoenicia, the village of Byblos, worshipped Tammuz, the dying and reviving vegetation god.

The Renaissance

The Egyptians worshipped him as Osiris, the god of resurrection, and his symbol was mummification. Natron had to be imported to the colony for the embalming procedure as well as for purification rituals. There would, then, have been a steady traffic of natron from Egypt to Phoenician ports for its use in the temples of Byblos and in outlying shrines.

The accidental discovery of glass on Phoenician shores, and its exploitation by natives, is certainly reasonable, and it is substantiated by Egyptian sources that first document the evidence of glassmaking on the Phoenician coast. Glass appeared as part of the vast booty that the pharoahs of the Eighteenth Dynasty began to bring back from Syria and Lebanon after the capture of Megiddo by Thuthmose III in the fifteenth century B.C. The lists of items and their places of origin are engraved on the walls of the temple of Amon-Ra at Karnak, where Thuthmose deposited his booty.[3] With this conquest came a glass-manufacturing technique that enabled the Egyptians to experiment with glass for mirror making. The Phoenician, later Syrian, glassmakers followed the market into Egypt and later into Greece, Italy, and Gaul, laying the foundations of the later European glass industry.

Medieval Venice, with its great seaport and access to the Mediterranean, was probably the first European country to obtain the raw products of glass manufacture from Syria and to learn its manufacturing techniques. The Venetians were making and exporting glass well before the tenth century. At the same time, a glass industry was established in Northern Europe in the Seine-Rhine region by Syrians, who held a monopoly there until the beginning of the seventh century. The monopoly was broken, according to Jean Morin, by the growth of anti-Semitism in Merovingian Gaul during the late sixth and the early seventh century. This movement has usually been identified as a religious conflict between the Christian church and the Jews, but the real cause was racial and commercial.[4] In any event, glass-production techniques were absorbed by Germany and the Netherlands.

Despite this setback, Syria continued to be an important center of glass manufacture and a world supplier of the spe-

The Mirror and Man

cial "glistening sand" from the beach of the Belus. As late as the fourteenth century, Sir John Mandeville related that "men comen fro fer watre by shippers and by lande with cartes, to fetchen of that gravelle."[5]

Although Germany and the Netherlands were making glass at the same time as the Venetians, the Venetians after many years of trial and error developed a much clearer glass than was possible before. This glass, greenish in color, was of excellent ductility, and many intricate articles of great brilliance were executed.

The Venetian art of glassmaking was difficult. It took three generations to make a master. Of necessity, glassblowing was a family business where fabrication procedures were handed down from father to son. This information was jealously guarded, and great secrecy surrounded the process. As early as 1224 Venice had a glassblowers' guild, and in 1279 an elaborate guild system was set up. In 1291 the industry was moved to the island of Murano to protect it from prying eyes. So successful was it that the emigration of workers and even the exportation of scrap glass was prohibited under penalty of death. As a result, a new kind of mirror appeared. The Venetians made these mirrors by pressing a metallic leaf to the back of plates of ornamental glass. They were in great demand throughout Europe, and after 1317 were a major contribution to decoration.

By the first half of the fifteenth century, the demand for glass mirrors again increased, not for household objects but as precious possessions or novelties. This demand was met by Venice and to a lesser extent by Germany and the Netherlands. Nuremberg had a guild for glass mirrors as early as 1373. Germany, in particular, not only catered to the wealthy but made a great number of mirrors for the religious trade, an enterprise that pleased the devout, who, every seven years, took part in holy pilgrimages, or *Heiltumfahrten*, to designated shrines.

The most important shrine was in Aachen. There, at the Liebfrauen Cathedral, were displayed the four most sacred relics of Christianity: the Virgin's garment, the swaddling clothes of the Infant, the loin cloth of Christ, and the kerchief of St. John the Baptist. The holy nature of mirrors lay in the manner in which they were employed by the pilgrims. They would hold these mirrors up to the sacred relics to catch them in a reflection. When they returned to their villages, they exhibited their mirrors to friends and relatives, boasting that they had brought back physical evidence as well as the inspirational qualities of their pilgrimage because their mirrors had captured the reflection of the sacred scene. The woodcut reproduced as figure 17 represents the shrine where reliquaries are displayed (known as a *Heiltumweisung*) and described by the clergy to a large group of men, women, and children. Some look up and marvel at the relics while others pray before them. Two of the women and a child can be seen holding objects composed of a central part and a decorated circular frame. These have been identified as *Heiltum* mirrors by Heinrich Schwarz.[6]

The manufacture of these mirrors was undertaken by Johannes Gutenberg, the inventor of movable type. This is documented in a lawsuit that arose as a result of unexpected financial difficulties.[7] Gutenberg and his associates formed a business to exploit an Aachen pilgrimage scheduled for 1439. But to their dismay the pilgrimage was postponed to the following year. Because there were money problems the partners quarreled and the disagreement led to a lawsuit. Gutenberg won the suit. The testimony shows that he taught a new method of mirror making to his partners.

In the first half of the fifteenth century in Venice, a master glassmaker, Angelo Beroviero, had the most famous furnace in Murano. He left a son, Marino, as skillful as himself. With other members of the family he invented a transparent glass about 1460 that was called "crystall" to distinguish it from the colored or green-tinted glasses that had been made up to that time. This was the birth of modern glass, and it soon led to the creation of the mirror as we know it today.

This came about in 1507 when Andrea and Domenico d'Anzolo del Gallo, who knew or perhaps discovered for themselves the tin amalgam process and combined it with the new Venetian glass to produce mirrors far better than those ever seen before. They submitted a petition to the Venetian Council of Ten in which they said "that possessing the secret of making good and perfect mirrors of crystalline glass, a precious and singular thing unknown to the whole world," and wishing to make large profits for Venice, they asked that an exclusive privilege be granted to them in all of the territory of the republic for twenty-five years.[8]

As this privilege promised to be profitable to the republic, and possibly to assure it the means of further strengthening a monopoly on glassmaking, it was granted for twenty years. The success of the marriage of crystalline glass to the tin amalgam process was enormous. Mirror making became a dominant industry. In 1564 the republic had to separate it from the other forms of glassmaking to create a separate company. The *specchiai*, or mirror makers, became artisans of distinction and established their own guild in 1569.[9]

The success of the tin amalgam process, also known as mirror foiling, which dominated the manufacture of mirrors for over three-hundred years, lies in its ability to be applied to glass of almost any size. And, more important, it could be produced and applied without heat. Tin and mercury were used at room temperature rather than as molten materials, thereby avoiding all the difficulties connected with the old method, like glass cracking, careful temperature control, and the limitation to small sizes.

The new process was not difficult. Tin was hammered into thin sheets and spread out very smoothly. Mercury was poured over the tin sheet and then rubbed into it either by hand or with a hare's foot. When the tin was saturated, it was covered with paper. The glass for mirroring was wiped clean and laid over the paper. Workmen pressed it down with their left hands, and drew the paper out very carefully with their right hands. The glass, now in contact with the tin amalgam, had weights placed on top of it. In a short time the amalgam adhered tenaciously to the glass to provide a beautiful mirror surface.

Heiltumweisung *in Nuremburg (woodcut from the Nuremburg* Heiltumbuch, *Peter Vischer, 1487). (Bayerische Staatsbibliothek, Munich)*

The Mirror and Man

The way in which the glass sheets were made is important, for it was a fundamental step to a later, more sophisticated, process. Craftsmen gathered a quantity of molten glass on the end of a blowing iron, which was an iron tube. Then they stepped upon a stand about five feet high so that they could blow a glass bubble elongated to the shape of a sausage. When it was sufficiently large and the walls were of even thickness, the two ends were cut off and a slit was made along the length of the cylinder that remained. The sheet of molten glass was pulled apart and rolled flat. The glass was allowed to cool slowly and annealed, after which it was ground and polished. This was not suitable for the production of perfectly flat and even sheets of glass, but the craftsmen grew so skilled that they succeeded in spite of the clumsiness of the method. The size of their glass sheets was limited to about forty-five inches in length and thirty inches in width. Larger sheets would warp and distort the reflections produced, and they were often too thin to withstand grinding and polishing.

The Venetian mirror industry flourished beyond all expectations. Europe demanded all that could be made. To safeguard this lucrative business, a corporation was formed "to protect the secret rights of mirror foiling."[10] This monopoly was maintained for over one-hundred years, and every method, including terrorism, was used to prevent disclosures of the technique.

Along with development of the large looking glass during the sixteenth and seventeenth centuries, hand mirrors and pocket mirrors reached their height of popularity. Pocket mirrors were an essential part of dress, as shown in portraits of the beauties in those days, for example, those done by Hendrik Goltzius, the Dutch painter and engraver. These portraits nearly always show a mirror, square, octagonal, or round, in various settings, sometimes as a centerpiece of a fan, but more often attached to the girdle by a ribbon or chain. This is confirmed in a stage direction of Massinger's 1624 play *The City Madam*: "Enter Lady Frugal, Anne, Mary and Millicent in several affected postures with looking glasses in their girdles."[11] These mirrors usually were in an embossed frame about four inches high and two inches wide. Pocket mirrors were popular gifts in the highest circles of the court, but sometimes they were not given with the best of intentions. One little mirror of Italian origin, apparently given as a farewell gift by an escapee from a duke's "ugly elderly daughter," had the inscription "Make no complaint against me lady, I only render back what you give me."[12]

The use of the hand mirror for self-adulation and adornment was rampant. Men and women paid great attention to their complexions and used cosmetics liberally. This inflamed the moralists of that time, who considered acts of self-admiration or self-beautification in a mirror whether by a man or a woman, as vulgar acts befitting the likes of the devil, and if not a sin, surely a sign of the decay of the nation. Women bore the brunt of the complaints. Thomas Nashe refers to women who prepared to go "a banquetting" by "practising with theyr Looking-glasses, how to . . . glance and looke alluringly amiable."[13] The French moralist Jean des Caurres was appalled by the moral decay of the day. He observed that ladies carrying mirrors fixed to their waists kept their eyes in perpetual activity. This, he felt, would result in their eternal damnation. He complained, "Alas! what an age we live in to see such depravity as we see, that induces them even to bring into church these scandalous mirrors hanging about their waist! Let all histories—divine, human and profane—be consulted, never will it be found thus brought into public by the most meretricious of the sex. It is true at present none but the ladies of the court venture to wear them, but long it will not be before every citizen's daughter and every female servant will wear them."[14]

As a result of a reform movement, a new fashion arose in the middle of the sixteenth century. The pocket mirror was then disguised inside the cover of a little book that looked to the uninitiated like a prayer book. This fashion invaded the highest levels of the aristocracy. Anna, "Queen of the Romans," and consort of the future emperor Maximillian II, had a mirror concealed in a little book with a beautifully embroidered cover. Dandies in the court of King Henry III of France carried their mirrors in book forms, stylishly, in their right-hand breeches pockets.

These deceptions did not fool everyone. The poet Charles Fitz-Geffry, in his 1617 "Notes from Black Fryers," exposed the conceited dandy who

> *Never walkes without his looking glass*
> *In a tobacco-box or diall set,*
> *That he may privately conferre with it*
> *How his band jumpeth with his peccadilly*
> *Whether his band-strings ballance equally*
> *Which way his feather wags.*[15]

Moralists termed the mirror a flattering glass, an item that nourished the vanity of the vain. The term became so popular that it was often used metaphorically by the writers of the period. Heywood, in his *A Woman Killed with Kindness*, has a character remark: "His sweet content is like a flattering glass / To make my face seem fairer to mine eye."[16] The account of Queen Elizabeth's death, quoted by Nichols in his *Progresses of Queen Elizabeth*, notes that "in the melancholy of her sickness, she desired to see a true looking glass, which in twenty years she had not sene, but only such a one as was made of purpose to deceive her sight: which glasse, being brought her, she fell presently into exclayming against those which had so much commended her, and took it so offensively, that some which had flattered her, durst not come into her sight."[17]

Shakespeare, too, makes use of the flattering glass in *Richard II*, but he quickly turns it into a mirror for self-knowledge (the ideal purpose for the mirror). He has Richard II, at the moment of crisis, change his nature from conceit to humility. A successful insurrection has been led against Richard II as a result of his failure as a ruler. Dissolute and avaricious, he was encouraged by his sycophantic followers to enjoy the ceremonies and symbols of his station, but he was unwilling to assume the responsibilities of being a leader. Instead of being wise, brave, and just, he became the example of misconduct.

Obliged at last to deal with the realities of his desperate situation, he asks for a mirror to study the false image of himself:

> *Give me the glass, and therein will I read.*
> *No deeper wrinkles yet? Hath sorrow struck*
> *So many blows upon this face of mine*
> *And made no deeper wounds? O flattering glass,*
> *Like to my followers in prosperity,*
> *Thou dost beguile me! Was this face the face*
> *That every day under his household roof*
> *Did keep ten thousand men? Was this the face*
> *That, like the sun, did make beholders wink?*
> *Was this the face that faced so many follies,*
> *And was at last outfaced by Bolingbroke?*
> *A brittle glory shineth in this face—*
> *As brittle as the glory is the face,*
>   *(Dashes the glass against the ground)*
> *For there it is, cracked in a hundred shivers.*
> *Mark, silent King, the moral of this sport,*
> *How soon my sorrow hath destroyed my face.*[18]

Despite the implications of the flattering mirror, moralists fought a losing battle against man's new perception of himself and his world. The spiritual importance expressed by the ancient, dim, and imperfect Pauline mirror was being replaced by a worldly outlook exemplified by the newly discovered looking glass, which reflected nature clearly and accurately. For intellectuals and Renaissance poets and painters, this brought a change in their attitude toward the mirror.

Poets like Donne and Spenser rejected the Pauline conception—that what one saw was an imperfect representation of one's self. By reinterpreting it, they shifted the ultimate revelation promised by Paul, to the here and now, arguing that by seeing one's true appearance in the mirror one finds a full knowledge of the soul. Their rationale, as they expressed it in their writings, was that earthly knowledge is a mirror faithfully and clearly reflecting the glory of God.

Allusions to the Pauline mirror by these poets indicate that a theological connection was necessary to Renaissance interpretations of the nature of mind, knowledge, and the soul. That is, while the mirror might variously symbolize mind, human essence, understanding, or fancy each of which was indispensable to the shaping and development of human knowledge, these writers felt the need to develop a variety of plausible interpretations to imply adherence to Pauline doctrine. It would have been beyond their belief and understanding to disdain or dismiss God.

Dogmatists resisted this change. They felt that to illuminate the mirror is beyond man's power in this world and counter to the disposition of God. And they believed that any attempt to adjust this idea by the advance of human science was at best a futile occupation and, at worst, a deception that might lead to the imminent danger of hell.

Since the art of mirror making had resulted in products of very high quality that supported the Renaissance philosophy of reality, believers in the Pauline mirror philosophy specifically blamed that industry for the fearful turn humanity had taken. Henry Cornelius Agrippa of Nettesheim excoriated mirror polishers in his diatribe *The Vanity of Arts and Sciences* (1530):

> It is an old Opinion and the concurring and unanimous judgment almost of all Philosophers, whereby they upheld, that every science addeth so much of a sublime Nature to Man himself, according to the Capacity and Worth of every person, as many times enables them to translate themselves beyond the Limits of Humanity, even to the Celestial Seats of the Blessed. . . . But I, perswaded by reasons of other nature, do verily believe, that there is nothing more pernicious, nothing more destructive to the well-being of Men, or to the Salvation of our Souls, than the Arts and Sciences themselves.[19]

Despite such protests, writers of the Renaissance succeeded with new ideas by imposing the properties of the real mirror upon the spiritual mirror, thereby diminishing its enchantment. In the humanistic philosophy of the times, the mirror reflected clearly, and the clear mirror became the clear spirit, and the clear spirit then liberated man.

Of the English Renaissance poets, Milton felt most strongly that clear spirit. He believed that he was looking in a mirror clearly, and that he, the mortal poet, could know God as clearly as he was known by God. He believed that his clear spirit was attained by hard work and the study of nature, man, and moral philosophy. Gordon W. O'Brien, in his *Renaissance Poetics and the Problem of Power*, illuminates the nature of the Renaissance mirror metaphor when he writes:

> John Milton's understanding of himself as he aims to tell us when, in Lycidas, he speaks of a clear spirit is this: I, who have abjured worldly pleasure, by disciplined study and meditation have rectified the Adam in my nature, so that no longer do I see the nature of things in a glass darkly but in a glass clearly. My spirit is a mirror which reflects without distortion the shape of all things in the cosmos to my soul; it is a mirror larger than the world, for it shows me things of which this world knows nothing. It is a glass which promises me omniscience and persuades me, that I, too, am the son of God, the heir of absolute glory and absolute power. Only such as I are fitted to the poet's task—only we whose glass is cleared, who are potentially something better than the angels— can cast the image of eternal providence.[20]

While poets created metaphors with a weakened Pauline influence that incorporated the attributes of the new looking glass, painters, by way of mirror representation in their work, were moving from spiritual matters to reality. The mirror became increasingly important to them. Paintings, which had dealt with the sublimity of heaven, began to be replaced by scenes from the natural world.

The Mirror and Man

In the early phases of the period, the artist did this by joining spiritual meaning with reality. The outstanding example of simultaneously representing religious symbolism and reality is in a painting where a mirror is prominent. This is the portrait of Giovanni Arnolfini and his wife, painted by Jan van Eyck in 1434 (fig. 18) This work is considered to be one of the earliest purely realistic representations of people in their everyday surroundings. It appears to be the detailed representation of a well-to-do Flemish dwelling at the beginning of the fifteenth century. But Erwin Panofsky has shown that the painting superimposes a spiritual meaning on a tranquil domestic scene. The painting is a kind of pictorial marriage document, and the room, with its symbolic and religious implications, represents a nuptial chamber. For example, the chandelier is a symbol of the Virgin. Its one burning candle—the "marriage candle"—is a symbol of the presence of God and his pervasive wisdom, and the dog at the feet of the couple is not just a household pet but a symbol of fidelity.[21]

The mirror, however, holds the viewer's attention because of its position on the vertical axis of the painting between the chandelier and the joined hands of the man and woman concluding the marriage ceremony. This large convex glass mirror reflects not only the reverse of the scene facing the spectator but also two other figures, one of them undoubtedly the artist himself. The mirror seems to appear as a handsome accessory in a intimately furnished room, but its frame is adorned with ten circular representations from the life and death of Christ—starting with the Mount of Olives and ending with the Resurrection—a clear indication that the mirror, too, is of symbolic significance. It might be interpreted as an allusion to the Holy Virgin and to the redemption of the world through Christ's incarnation and death. At the same time, a mirror that verifies the sacramental ceremony is the symbol of the all-seeing eye of God.

Another interesting example of this kind is the 1449 St. Eligius painting of Petrus Christus (fig. 19). The scene is a natural one. The goldsmith sits behind his workbench,

Jan van Eyck, Marriage of Arnolfini. *(National Gallery, London)*

weighing a golden ring for a bridal couple. In the foreground on the right side of the table stands a convex mirror. It is positioned to expand the scene to include the outside so that it catches a view of the street in a Flemish town where, at this moment, another couple seems to approach the workshop. According to Heinrich Schwarz, the mirror must have also had a religious significance related to the bride's virginity. The mirror reflects the fleeting image of the world and may, therefore, also be a symbol of vanity pointing to the transitoriness of the delusive treasure surrounding the goldsmith. Mirrors and their assumed magic powers were used for the detection of thieves, and the mirror on the counter of the goldsmith may also have been a device against theft.[22]

These two paintings represent the transition that was taking place in the perception of reality, but the mirror in the Arnolfini portrait may be considered the first significant example of the fusion between the contending forces of the past and the future, the world of symbol and the world of visible fact. In time the metaphysical outlook was to fade and yield to an objective attitude, one lacking any symbolic or religious meaning. The exact copying of nature became a vogue, and the mirror with its qualities of reflection was to become a most important artist's implement.

Even before Jan van Eyck's time the mirror had entered the artist's studio, primarily as an indispensable aid to the artist painting his own portrait. An early example is found in a French illuminated manuscript of Boccaccio's *De claris mulieribus*, finished in 1402: it is a miniature of the nun Marcia painting her own portrait with the aid of a mirror. The mirror is convex, and, as pointed out earlier, it was the mirror easiest to make in those days.

The frequent appearance of convex mirrors and their reflections in early Flemish art, may be taken as another proof of the growing spread of the use of mirrors by artists.

The extent of the aid of the convex mirror to artists can be seen in many fifteenth-century paintings. Of specific interest is a painting attributed to Conrad Witz entitled *Holy Family with St. Catherine and St. Barbara* (c. 1440). In this work the

Petrus Christus, Saint Eligius. *(The Metropolitan Museum of Art, Robert Lehman Collection, 1975. All rights reserved, The Metropolitan Museum of Art)*

pillars of the cathedral are bowed at the bottom and top. The cause of this distortion is most likely due to the use of a convex mirror from which the artist painted the reflection.

The flat, or plane, mirror was also beginning to take its place in the artist's studio in the fifteenth century. In the first modern treatise on the theory of painting, *Trattato della Pittura*, (1435–36), Leon Battista Alberti comments on the importance of the mirror to the artist: "A good judge for you is the mirror. I do not know why painted things have so much grace in a mirror. It is marvellous how every weakness in a painting is so manifestly deformed in the mirror. Therefore, things taken from nature are corrected with a mirror. I have here truly recounted things which I have learned from nature."[23]

A few years after Alberti, Filarete recorded similar experiences. Later, Leonardo elaborated on Alberti's and Filarete's observations. In 1492 he wrote his famous "How the Mirror Is the Master and Guide of Painters":

> When you wish to see whether the general effect of your picture corresponds with that of the object represented after nature, take a mirror and set it so that it reflects the actual things, and then compare the reflection with your picture, and consider carefully whether the subject of the two images is in conformity with both, studying especially the mirror. The mirror ought to be taken as a guide—that is; the flat mirror—for within its surface substances have many points of resemblance to a picture; namely, that you see the picture made upon one plane showing things which appear in relief, and the mirror upon one plane does the same, the picture is one single surface, and the mirror is the same."[24]

Because of the importance of the mirror to the artist, painters and mirror makers were closely connected in the fifteenth century and, as in Bruges, even united in the Guild of St. Luke, the patron saint of both professions. Thus a close link connected the painter, who sought to attain an image as precise as only the mirror (the symbol of truth) could yield, and the mirror maker, who could provide the artist with something that could reflect and retain such an exact image of reality.

Soon plane mirrors became an invaluable tool for the artist. As Leonardo noted, it provided objectivity by "seeing" objects reversed from right to left, but it also showed what, according to the laws of linear perspective, would not be visible: that is, it made it possible to view an object from different angles not in the line of vision of the artist.[25]

Linear perspective, one of the greatest advances in the art of painting, provided a satisfying illusion of depth or of three dimensions in the faithful representation of real objects. The understanding and application of linear perspective was made possible by the use of the plane mirror. This discovery is attributed to the architect Filippo Brunelleschi in 1425. This was noted by Filarete in his treatise on architecture, written during the early 1460s: "And so I believe that . . . [Brunelleschi] the Florentine found the way to make this plan [linear perspective] which truly was a subtle and beautiful thing, which he discovered through considering what a mirror shows to you."[26] Brunelleschi was seeking a design for the dome of the unfinished cathedral in the Piazza del Duomo of Florence. He saw in the effect of perspective a way to design a dome of proper proportions. He investigated this idea with an experiment that is described by Samuel Edgerton, Jr., from a detailed account by Antonio di Tuccio Manetti, the scholar, mathematician, and biographer of Brunelleschi.[27]

Brunelleschi's experiment was carried out at the piazza between the unfinished cathedral and its facing Baptistery. He used a twelve-inch-square mirror and a wooden panel of the same dimensions. He set up his mirror in the central entrance of the cathedral to reflect the Bapistery and its surroundings. He planned to use the reflection as a model to be painted on the panel, which rested on an easel beside the mirror. Brunelleschi determined the vanishing point of the reflected Baptistery by a dot on the mirror and placed a dot at the same location on the panel and then proceeded to copy the mirror reflection. When he completed the painting, he proceeded with his experiment.

"The artist then drilled a little hole through the back of the completed painting" at the vanishing point. Holding the panel up against his face, "close to one eye, with the painted surface facing away from" him, he "peered through the little hole . . . at the *reflection* of the picture in a mirror, held with his other hand." He could see that all the lines of the piazza and the architectural setting, which he had painted as extending into the distance around the Baptistery, converged to a point identical with his eyepoint. This was scientific proof of the principle of the vanishing point.[28]

Brunelleschi demonstrated his findings in a spectacular way to his colleagues. He brought visitors to the spot where he had painted his picture and, with them facing the cathedral, asked them to hold the panel against one eye so that they could look through the hole from the back. Brunelleschi placed the mirror in the viewer's other hand to reflect the scene so that it completely filled the mirror. In this panel the sky was not painted; instead, its border was silvered and polished, so that it, in turn, reflected the actual sky and passing clouds. This gave the scene the appearance of real life. With the paraphernalia in hand, the viewer would be turned around so he faced the Baptistery. Then, with his viewer entranced, Brunelleschi would suddenly remove the mirror from the hand of his guest so that the real Baptistery appeared before him through the hole. For a moment he would believe he was still viewing the reflected painting, so alike were the two.

With this ingenious technique, Brunelleschi was able to design several domes and "try them out," as it were, on top of the unfinished cathedral before choosing the most suitable one. Eleven years later a dome of outstanding beauty was finally constructed from one of his designs.

Dutch artists also have a tradition of using the mirror as an aid to their art. Gerard Dou was a seventeenth-century painter who studied with Rembrandt and succeeded him as a leading artist in Leiden. His highly finished works have a mirrorlike quality that may not be just a coincidence. To him is attributed an invention

> which reduces large objects to a small scale. Dou used a kind of screen fixed to his foot; in the screen he had inserted and framed a [convex] glass at the height of his eye when he was seated at his easel. This screen formed a species of partition between the object to be represented and the artist. The object was seen in a reduced scale in the glass and the artist had only to copy its form and colour. . . . Dou then drew the objects on to his canvas which was divided into equal squares corresponding to the threads on a little frame of the exact size of the circumference of the glass, in such a way that when the frame was attached to the glass it represented a square drawn within a circle.[29]

Dou was only one of many Dutch artists of the day who used mirrors to obtain a true imitation of nature. Distinguished painters like Carel Fabritius, Pieter de Hooch, Hoogstraten, and Janssens also made use of them in their work. Of particular interest is Jan Vermeer, the outstanding artist of the period. He made frequent use of the mirror to assure the fidelity of the architectural detail of his pictures. In his search for a true representation of the human form, furniture, or architectural features, he was conscious of the optical problem involved in proper perspective and spatial placement. By painting scenes reflected from one or two mirrors, he attained an objectivity almost impossible to obtain by a direct view of a scene.

One of Vermeer's famous paintings, *The Love Letter* (fig. 20), shows two women in an adjacent room seen through a doorway. It appears that the curtain and chair in the foreground are in the room where the artist was working. According to R. H. Wilenski in his *Dutch Painting*, the doorway was a mirror reflecting the models in another part of the room in which Vermeer was working. Wilenski further suggests that the curtain hanging from the mirror with the chair

Jan Vermeer, The Love Letter. *(Rijksmuseum, Amsterdam)*

next to the mirror gives the illusion of a doorway, especially as the artist viewed it:

> Vermeer sat with his back to the curtain, chair and mirror, looking into another mirror by the side of his canvas. Reflected in this second mirror, the curtain, chair and mirror became the foreground and the rest of the picture is the reflection in the back mirror. Vermeer sat a little to the left of the chair. The models were . . . possibly in front and to the right of him. The broom which appears to be leaning across the entrance to the second room was probably substituted at the last minute for the leg of the artist's easel which was reflected in the back mirror at this point.[30]

In another example, *The Painter and His Model* (fig. *21*), it seems likely that Vermeer used the same technique. The back mirror reflected the artist's back and easel and his model in front of him (against the wall), and he painted all this as re-reflected in the front mirror. The foreground was a reflection in the front mirror of the curtain surrounding the back mirror and the chair.

Artists in other countries also used mirrors during this period as an aid to their painting. The most famous perhaps was Velasquez in Spain who, like Vermeer, used mirrors to produce some of his masterpieces, such as *Las Meninas, The House of Martha*, and *Venus and Amor*. Confirmation of his use of mirrors can be found in the inventory taken after Velasquez's death. It lists ten mirrors, a goodly number to possess in those days.

The mirror gave painters an advantage over sculptors in the presentation of the human form. Not only did it permit an object to be viewed from different angles, it also presented these views simultaneously. The toilette of Venus became for painters in the sixteenth and seventeenth centuries a favorite theme. Velasquez, as well as Tintoretto and Rubens, show in their treatment of this subject the nude body from the back, but a conveniently held mirror shows at the same time the face, both to Venus herself and to the spectator.

Jan Vermeer, The Painter and His Model.
(Kunsthistorisches Museum, Vienna)

The Renaissance

A conveniently held mirror could also be used in an unscrupulous manner. Figure 22 shows a method used to cheat at cards. Here an accomplice holds a mirror to reflect the contents of the player's hand to her confederate.

Also during the sixteenth century opticians invented a technique for reconstituting distorted pictures by viewing their reflections from a convex cylindrical mirror. In these representations the original scene is deliberately distorted according to a specified procedure that makes the painting, or a part of it, so misshapen that the eye cannot trace in it a resemblance to any regular figure. Only when viewed on the convex surface of a cylindrical mirror resting like a chimney on top of the horizontally placed distorted painting do the figures appear in a normal fashion. This technique, known as anamorphism, was used to disguise malicious humor poked at royalty or to portray explicit pornographic scenes.

The century also saw the final unraveling of the mirror secrets of the ancient magician-priests. Culminating the work of Heron and Bacon, the Italian physicist Giovanni Battista della Porta revealed that this sorcery was only a manipulation

Cheating at Cards. (Mansell Collection, London)

The Mirror and Man

of optical laws. A large part of his comprehensive book *Natural Magick*, published in 1558, deals with the science of mirrors and the things that may be done with that knowledge.

The invention of the looking glass literally and figuratively contributed to the Age of Humanism, an era of cultural change that substituted a worldly point of view for a religious one in art, literature, and government. Man became the point of reference rather than God. The clear and truthful looking glass literally showed man reality, and this led him to abandon the "dark glass" of Pauline philosophy. In the spirit of the times society embraced the virtues of the looking glass. Or is it possible that the spirit of the times was influenced by the looking glass? It is thought-provoking to speculate: if the clear and true mirror had not been invented when it was, would the fruition of this new age have been diminished?

Evidence suggests that it would have. As the Age of Humanism evolved, with its recognition of the here and now, the metaphor of the Pauline mirror created a theological-secular conflict. This conflict resulted from man's clouded view of himself and his spirituality clashing with a new view of himself in the world. This ambivalence would have slowed the flowering of the Renaissance if the well-established Pauline mirror had not been challenged by a secular counterpart. The new true and clear mirror that accurately reflected man and his world helped resolve man's attitude toward reality. Metaphorically, he could in his lifetime know God as clearly as he was known by God.

NOTES AND SOURCES

1. Pliny, *Natural History*, 10 vols., tr. D. E. Eicholtz, Loeb Classical Library (Cambridge: Harvard Univ. Press, 1938–63), 10:190.

2. Ibid.

3. Anita Engle, *Readings in Glass History* (Jerusalem: Phoenix, 1973), p. 5.

4. Raymond McGrath and A. C. Frost, *Glass in Architecture and Decoration* (London: Architectural Press, 1961), p. 31.

5. Ibid.

6. Heinrich Schwarz, "The Mirror of the Artist and the Mirror of the Devout," *Studies in the History of Art for the Samuel H. Kress Foundation* (New York: Phaidon Press, 1959), p. 103.

7. Douglas C. McCurtrie, *The Gutenberg Documents* (New York: Oxford Univ. Press, 1941), pp. 100–123.

8. Alexandre Sauzay, *Wonders of Glass Making* (New York: Charles Scribner, 1870), p. 95.

9. McGrath and Frost, *Glass*, p. 32.

10. Frances Rogers and Alice Beard, 5000 Years of Glass (Philadelphia: Lippincott, 1948), p. 197.

11. Act 1, scene 1.

12. Max von Boehn, *Modes and Manners* (New York: Dutton, 1929), p. 202.

13. Thomas Nashe, *Christ's Teares over Jerusalem* (1593; rpt. ed. Menston, England: Scolar Press, 1970) p. 70.

14. Jean des Caurres, Recueil des Oeuvres Morales et Diversifiées, 1575, in Horace Greeley et al., *The Great Industries of the United States*, 2 vols. (Hartford: J. B. Burr & Hyde, 1872), 2:767.

15. S. H. Cuming, "On Mirrors," *Journal of the British Archaeological Association*, 1861, p. 286.

16. Act 1, scene 1, lines 33–34.

17. John Nichols, *The Progresses and Public Processions of Queen Elizabeth*, 3 vols. (London, 1823; rpt. ed. New York: Burt Franklin, 1966), 3:612.

18. Act 4, scene 1, lines 276–91.

19. Quoted in Gordon W. O'Brien, *Renaissance Poetics and the Problem of Power* (Chicago: Institute of Elizabethan Studies, 1956), p. 18.

20. Ibid., p. 48.

21. Heinrich Schwarz, "The Mirror in Art," *Art Quarterly* 15, no. 2(1952):97.

22. Ibid., p. 103.

23. Leon Battista Alberti, *On Painting*, tr. John R. Spencer (New Haven: Yale Univ. Press, 1956), p. 83.

24. Edward MacCurdy, *The Notebooks of Leonardo Da Vinci* (New York: Reynal and Hitchcock, 1938), p. 879.

25. Ibid., p. 888.

26. Quoted in Samuel Y. Edgerton, Jr., *The Renaissance Rediscovery of Linear Perspective* (New York: Basic Books, 1975), p. 125.

27. Ibid., pp. 124–52.

28. Ibid., p. 127.

29. Reginald H. Wilenski, *Dutch Painting* (New York: Beechhurst Press, 1955), p. 188.

30. Ibid., p. 189.

The Mirror and Man

Other sources for this chapter include: Jurgis Baltrusaitis, *Anamorphic Art*, tr. W. J. Strachan (Cambridge: Chadyck Healey, 1977); Margaret MacDonald-Taylor, "Mirrors of High Fashion," *Country Life* 136 (1964):970; R. W. Symonds, "English Looking-Glass Plates and their Manufacture," *Connoisseur* 97 (1936): 243–48; Peter Ure, "The Looking-Glass of Richard II," *Philological Quarterly* 34 (1955): 219–23; Geoffrey Wills, "From Polished Metal to Looking-Glass," *Country Life* 124 (1958):939–43; Geoffrey Wills, *English Looking-glasses* (South Brunswick, N. J.: A. S. Barnes, 1965): Charles Yriarte, *Venice*, tr. F. J. Sitwell (London: G. Bell, 1880); Wolfgang M. Zucker, "Reflections on Reflections," *Journal of Aesthetics and Art Criticism* 20 (1962):239–50.

THE SEVENTEENTH
AND THE
EIGHTEENTH
CENTURIES

THE MONOPOLY of the Venetian mirror industry was finally broken in a ruthless manner by Colbert, the French controller of finances under Louis XIV who strengthened and brought prosperity to his nation. He created the French navy and merchant marine, chartered trading companies, established colonies in Canada, and developed silk, woolen, and glass industries.

When Colbert came into power, he was aware of Venice's almost complete control over the manufacture of mirrors. Also he knew that unsuccessful efforts had been made by Louis XIV and his three predecessors to enter this lucrative market. He was appalled when he learned that France each year spent exorbitant sums for Venetian mirrors to supply fashionable people with articles for the decoration of their boudoirs and reception rooms and by the king for the embellishment of his royal dwellings. He decided that the manufacture of mirrors must be introduced into France by whatever means possible so that the country could share in these enormous profits.

In the fall of 1664 Colbert wrote to Bonzi, the French ambassador to Venice, to tell him to enlist the services of some Venetian mirror makers and to send them to France as soon as possible. Bonzi quickly replied that his first efforts were unsuccessful. He reported that the workers of Murano, of whom he inquired, were fearful of the offer. They were certain that if they emigrated their families would be subjected to retaliation. In spite of these obstacles, Bonzi said he would try again.

Colbert waited impatiently until the end of the year. Then he ordered Bonzi to obtain workers by any means possible. The ambassador then began unauthorized negotiations with a group of mirror makers, and five months later Colbert sent an agent, the sieur de Jouan, to Venice to assist him. Jouan was put in touch with a mirror maker named La Motta, who for a substantial sum of money agreed to go to France. Toward the end of June 1665 Jouan succeeded in smuggling La Motta and three of his assistants out of Venice. The Venetian authorities soon uncovered the plot and threatened the relatives of the defectors with punishment. They directed their ambassador at Paris to persuade the fugitives to return at once.

After an intensive investigation Ambassador Sagredo failed to locate La Motta and his aides. Meanwhile, Colbert did not rest. He bribed Castellan, a known Italian glassmaker, with a 4,600-livre expense account and arranged for another group of workers to be smuggled out of Venice. Castellan sent his nephew on this covert mission, and after some narrow escapes he arrived back in France in July 1665 with ten mirror makers. Four panicked at once and were allowed to return home. The remaining six were brought to Paris, but Sagredo discovered them and threatened them with dire consequences if they remained there to work. These fears were overcome by gifts and favors from Colbert. Their leader, Antonio de la Rivetta, was given a pension of 1,200 livres a year in addition to other benefits, and the other workers were given smaller pensions.

Before the end of the year, still another batch of Venetian workers arrived. Upset and angered, the Venetian authorities ordered Ambassador Sagredo to use any means in his power to compel the workers to return to Venice. He was specifically given the power to offer them safe-conducts and the promise of employment. He failed. This led to his replacement by Ambassador Giustiniani, who had a little more luck. He persuaded the workers to promise to discontinue teaching their secrets to anyone, and in April 1666 he was able to send back to Venice a few of the less important artisans who had not fared so well financially.

Colbert counterattacked. He tried to get the wives of the workers to join their husbands. Then he greatly increased the pensions of some of the artisans who agreed to train French apprentices. As a supreme gesture, Colbert induced Louis XIV to visit the mirror works with him on April 29, 1666. The king showed great interest in these operations and the skill of the Italians, and he asked Colbert to make them presents of money on the spot.

This last touch was too much for the Venetian authorities. They asked Giustiniani to promise the workers free pardons and gifts of money. They forged letters purporting to come from the wives of Rivetta and another mirror master, Civrano, demanding the return of the husbands. This was discovered by two men who realized that the letters had been prepared by persons more intelligent and better educated than the deserted wives.

Colbert, aware of all this, outwitted the Venetian authorities by having a forged letter delivered by a special agent to Murano. He brought the wives of Rivetta and Civrano back to France. This was the limit for the Venetian officials. Giustiniani arranged for the assassination of Rivetta and other key workers. The ambassador, taking advantage of dissension at the factory, faubourg Saint-Antoine, got in touch with La Motta, who had been one of the first workers to come to Paris and who felt he had been less favorably treated than Rivetta. Urged on by Giustiniani, La Motta gathered an armed band and attacked his rival. But Rivetta, forewarned, collected his friends and pitched battle took place. La Motta and two of his supporters were wounded, and it ended only when a company of soldiers intervened.

Peace returned to the plant but not for Giustiniani, who continued his intrigue. A short time later two of the best workers at the mirror factory died violently. No doubt they were poisoned.

The foresighted Colbert had laid the groundwork for a mirror-manufacturing business in France by forming a company in January 1665 under the aggressive management of Nicholas du Noyer, a man who had had some experience in glassmaking. Associated with him was a group of officials who provided financial support and a willingness to cooperate. Further, Colbert secured for Noyer 12,000 livres from royal funds to aid him with initial expenses. Noyer set up a factory in faubourg Saint-Antoine, and when the first Venetians came later that year they were put straight to work.

During the following months, Noyer, fully aware of the intrigue underway, was concerned with the stability of the company. He reported to Colbert that while the company was able to make satisfactory mirrors, its condition was precarious. The Venetian workers imparted their skill grudgingly. If they were to die or leave, the company stood to lose most of its investment. Yet Noyer thought too much had been invested to drop the endeavor, and he knew that the secrets of the Venetians had to be wrested from them at any price. He proposed to Colbert that the workers be given an estate worth 60,000 livres, that they be promised pensions for their wives and children, and that they be given 2,000 livres for each of the French apprentices they trained. He was doubtful, however, that even under such extravagant terms they would accept. If they refused, Noyer suggested that a new plant be set up and that all Frenchmen who claimed to know anything about making mirrors be invited to come there and carry on experiments.

Noyer's concern proved correct. The armed outbreak and poisonings greatly disturbed the Venetian émigrés. The chief workers decided to go home, but they were in a position to drive a hard bargain. They demanded and received a

The Seventeenth and Eighteenth Centuries

written pardon and a gift of 5,000 ducats. Giustiniani negoti-
ated the arrangement, and the Venetian authorities were
very pleased with this settlement, which they fully honored.
Later, some of the Italian workers who missed the good life
in Paris wanted to return, but Colbert had had enough. He
felt he had got all he could out of the Venetians and he
turned them down. The French were now making satisfactory mirrors without the Venetians. For their efforts, Noyer
and his associates received an exclusive privilege for twenty
years from Louis XIV to have the right to make Venetian
mirrors, window glass, beads for the Indies, vases, enamel-
ware, and tableware, "both to serve for the ornament of our
Royal Houses and for the public convenience."

During this twenty-year exclusive period, no one in
France was to make these products without permission from
the company. The company was to have the right to secure
necessary materials from anywhere in France. Its products
were not to be taxed within France, and, on goods exported,
it was to pay one-half the regular duties. Noyer and his
associates were to have the privileges of members of the royal
household, reminiscent of the designer-molder-polisher trinity at the Japanese imperial court, and they were exempt
from all taxes and duties. The company was authorized to
use the inscription "Manufacture Royale des Glaces de Mir-
oirs," and its porters were allowed to wear royal liveries.

Within five years the company was manufacturing mirrors with great success. By 1671 few mirrors were any longer
imported from Venice and all the king's needs were supplied
by Noyer's company. A true devotee, Louis XIV set the
fashion for mirrors. He purchased 700 of them in 1672. Apparently he was the first monarch to install mirrors in a royal
coach, and, of course, he was responsible for the mirror
panels set down the length of the famous Galerie de Glaces
at the Palace of Versailles.

By 1680 France had devastated Venice's mirror-manufacturing monopoly. Its exports deprived that country of
about one million livres a year. Of all the enterprises founded
by Colbert, the mirror industry was destined to survive un-

broken to this day. It was reorganized and had privileges renewed several times before the Revolution. In 1830 it was modernized and, as the Compagnie de Saint Gobain, it still exists as an important enterprise in France.[1]

A contributing factor to the success of the French looking glass industry was the discovery of casting glass as flat plates by Bernard Perrot in 1673. Perrot received a royal patent for his invention. The patent declared that he had discovered how "by a means hitherto unknown" to "cast glass in plates as is done in metals."[2] Up to that time, to make plates of glass for mirrors, it had been necessary first to blow the glass into cylinders, then to flatten and straighten the cylinders, a process that limited the size of flat mirrors.

France further improved the process by the invention of Abraham Thevart in 1688 for casting mirrors of a much larger size than ever before.[3] His first plates astonished the Parisians. They were 84 inches in height and 50 inches in breadth, exceeding even the Perrot plates. His casting method consisted of pouring molten glass quickly onto a flat table with raised edges and rolling it out to cover the entire surface to a uniform thickness. The glass was annealed, ground, and polished. France, now in possession of the Venetian foiling process and the new method of glass casting, produced excellent mirrors. For the first time in history glass was free from bubbles and flat as a pool of untroubled water. When backed with foil, the plate glass made a mirror never before experienced. Now France enjoyed the monopoly of the mirror market.

During the period of Colbert's great success in making France the leader in the art of mirror making, Charles II was restored to the throne of England. This brought about, in the words of Evelyn, the diarist, a "politer way of living."[4] Furniture and appointments in houses of the upper classes became more luxurious than before. The growth of the popularity of the looking glass was one of the many signs of this change. England was rebounding from the restrictions of the Puritans, and the repressed desires of the wealthy and not so wealthy were awakened. This resulted in their indulging

The Seventeenth and Eighteenth Centuries

vanities long forbidden. Charles II led this revival by purchasing many luxuries, including a special treat for his favorite mistress, Nell Gwyn. He had her bedroom paneled entirely in mirrors.[5]

When Charles II came into power, he was favored with a small, growing looking glass industry that had been established by Sir Robert Mansell about 1620. Preparing himself for his endeavor, Sir Robert brought over several Venetian glassmakers (not mirror makers) in 1611 to instruct the English in that craft. Soon he was making the first English mirrors but *without* the Venetian foiling process. The mirrors he made were small and gave a poor reflection. It was difficult to find a good English mirror then. In 1639 the mother of Edward Hurley wrote, "Dear Ned, if theare be any good looking glasses in Oxford, chuse me one . . . because I think you will chuse me one, that will make one a true ansure to onse face."[6]

No English glass factory before the Restoration was able to produce looking glass plates similar to those made by the Venetians. The Englishman who required a large looking glass and could afford to pay the price had to get one from Venice. The cost of a looking glass of modest size in 1664 is to be found in Pepys's *Diary:* "By coach with my Wife and bought a looking glasse by the old Exchange which costs me £ 5 5s. . . . A very faire glasse."[7] This would be the equivalent of several hundred dollars today.

During the same year, which coincided with Colbert's start, George Villiers, the second duke of Buckingham, appeared upon the scene. The duke, who was a gentleman of the bedchamber to Charles II and a privy councellor, was also an astute businessman. Sensing the mood of the times and Charles II's love of opulence, he had no trouble in obtaining a patent to make glass. The crafty duke had previously gained control of a number of patents held in the names of others and formed The Worshipful Company of Glass-sellers and Looking Glass Makers. In 1673 he established the famous glassworks at Vauxhall, which bore his name. By this time the duke, like Colbert, had obtained a number of Venetian mirror makers who were the initial craftsmen in this undertaking. Vauxhall specialized in the making of plate glass, which was used for looking glasses, coach glasses, and windows. This glass was unquestionably of a better quality than any previously made in England. The Venetian foiling technique was used, but the English were to continue to retain the blown-glass technique for another hundred years. They were able to blow plates to a size of 82 inches by 48 inches, and the quality of these large, blown plates, when made by skilled craftsmen, were presumably equal to the French cast plates, but they were rarely used. The large French plates could be imported and sold at a cheaper price despite the heavy import duty.

As the seventeenth century drew to a close, larger and larger looking glasses of good planar quality and clarity were being made. Mirrors were no longer only useful pieces of furniture. They also began to be used for aesthetic and architectural purposes. This was particularly true in England and France, where the artistic framing of mirrors enhanced their beauty. The French mirror craftsmen catered to the aristocracy with products that were elegantly regal, while the English craftsmen contributed immeasurably to the grace and decorative use of the mirror in the home.

As early as 1690 the English introduced the large overmantel mirror hung above the fireplace. This added spaciousness to the room. At the same time they began to place mirrors known as pier glasses against the walls comprising the solid uprights of masonry between windows. These were a great advance in interior design, for the mirrors reflected whoever or whatever stood in the light of the windows, and they brightened and enlarged the apartment. About this time the swinging toilet mirror on a box stand appeared. The mirror was enclosed in a frame that was attached to two uprights by swivel screws. The uprights were secured to a stand that often resembled a miniature bureau.

A new fashion of the period, about 1690, was the use of the mirror at one's reception toilet.[8] Evelyn saw the duchess of Portsmouth seated before a mirror having her hair combed

The Seventeenth and Eighteenth Centuries

by her maids while the king and a number of gentlemen stood near. A picture of Mary II shows the queen seated, arranging her hair in front of a dressing-table mirror. She wears the new high headdress called a *fontange*, and, significantly, the shape of many of the dressing mirrors at the turn of the century was tall with an arched top to fit the contour of the *fontange*.

John Gumley was an important entrepreneur in the looking glass trade. He opened an establishment in 1695 that sold looking glasses of great beauty to the well-to-do. His prominence and the pride the Englishman had in this new manufacture were expressed glowingly by Richard Steele in the May 13, 1715, issue of *The Lover:*

> I shall now give an Account of my passing yesterday Morning, an Hour before Dinner, in a Place where People may go and be very well entertained, whether they have, or have not, a good taste. They will certainly be well pleased, for they will have unavoidable Opportunities of seeing what they most like, in the most various and agreeable Shapes and Positions, I mean their dear selves. The Place I am going to mention is Mr. Gumley's Glass Gallery over the New Exchange. I little thought I should ever in "The Lover" have occasion to talk such a thing as Trade; but when a Man walks in that Illustrious Room, and reflects what incredible Improvement our Artificers of England have made in Manufacture of Glass in thirty years time, and can suppose such an Alteration of our Affairs in other parts of Commerce, it is demonstrable that the Nations who are possessed of Mines of Gold are but Drudges to a People, whose Arts and Industry with other Advantages natural to us, make itself the Shop of the World. We are arrived at such Perfection in this Ware, of which I am speaking, that it is not in the power of any Potentate in Europe, to have so beautiful a Mirror as he may purchase here for a Trifle[9]

Through the efforts of Gumley and others the demand for dressing and ornamental looking glasses increased. Advertisements boasting excellent mirrors often appeared in the local newspapers. One such advertisement appeared in *The Post Man*, February 13, 1700: "Large looking-glass Plates, the like never made in England before for size and goodness are now made at the old Glass house at Foxhall, known by the name of the Duke of Buckingham House, where all people may be furnished with rough plates from the smallest sizes to those of six feet in length and proportional breadth, at reasonable prices."[10]

The reasonable price for a finished mirror was exorbitant. For example, during the reigns of Queen Anne (1702–14) and George I (1714–27) it was customary to place below the pier glass a pier table that matched the frame of the glass: a gilt frame had a gilt table to match; a walnut frame, a walnut table. If you had shopped at a place like Gumley's for such a combination, you would have seen a pier glass six feet by three and a half feet with a round top and a carved walnut frame for £150; the solid walnut table and matching stands cost only £15.

The most elegant looking glasses were produced between 1690 and 1730. They possess a style that has never been surpassed. Characteristic of the early eighteenth century are Queen Anne mirrors, which have an upright-arched or surpentine-arched frame, often surmounted with an elaborate openwork cresting. Some years later an early Georgian design was introduced. Known as an architectural wall mirror, it was distinctive in its refined treatment of a baroque style. The frame was made of walnut or mahogany and was partially gilded. The upright frame, with its gilded moldings carved with classic ornament, was surmounted by a pendant. Fine oil paintings on canvas or glass were often combined with elaborate mirrors and hung as ornate overmantel mirrors. In the late Georgian period the illusory effect of the mirror was employed to advantage to extend the apparent size of small narrow hallways in apartments. Placed at the extreme ends of the hall, they presented the "appearance of an endless suite, a very grand piece of scenery."[11] Further, in those days of candlelight illumination, it was also desirable that mirrors should reflect and redouble the light of the chandelier or sconce. A few years later the French style known as Louis XV invaded England in the form of carved and gilt wood wall-mirrors in the rococo style.

Such famous interior decorators as Charles Cressant, Robert de Cotte, Jean François Blondell, and J. F. Leler contributed to the style of the French mirror in the eighteenth century. The French perhaps were not so interested as the English in the interior decoration of the house, but they did make significant contributions with their expertise with cast glass. Vista mirrors, or what the French call *glaces à répétition*, are two mirrors placed face to face. A beautiful example can be found in the Palais de Compiègne, where vista mirrors produce and infinite prolongation of the room's perspective. To achieve such a result without distortion of the reflecting images requires not only a truly flat surface but a parallelism of the two faces of the glass. Vista mirrors must be mounted in a way that will prevent the slightest deformation of their shapes. The glass must be colorless; otherwise the rapid darkening of the repeating images will prevent the illusion of a distant vista. In general, the French have always made masterful use of mirrors in architectural design. Perhaps because they invented cast glass, the French have maintained a superior initiative in the use of mirrors in public places.

It was not until 1773, when England through the newly formed British Cast Plate Glass Manufacturers successfully introduced the large-scale manufacture of plate glass by casting so that they could compete with and ultimately do without French imported glass. During the last quarter of the eighteenth century, large, bright, distortion-free mirrors were produced. This led to a variety of mirrors for home use that were long enough to reflect a person flatteringly in full length. A popular style was the cheval glass, a life-sized mirror enclosed in a rectangular frame that was attached to upright supports by swivel screws. The uprights rested on splayed bridge feet.

As a further result of the improved, less expensive plate

glass, another architectural innovation came into vogue, that of enlivening a room by reflecting the outside world. It was an inspiration to the poet Samuel Rogers, whose lodgings overlooked a fine view of the Thames. He covered his window shutters with mirrors so that from every part of the room he could see a broad expanse of the river and the moving craft upon it.[12]

By the close of the eighteenth century, the convex mirror, which had declined in popularity with the success of the plane mirror, came back into fashion. Known as girandole mirrors—a circular convex mirror with candelabra fixed before it—they came into use in France during the reign of Louis XVI. Soon they were adopted in England without the candelabra and became quite the rage. These mirrors were usually hung at the end of a room so they could reflect the whole of the interior. The compressed, delicately distorted scene produced, accompanied by the intensification of colors, lights, and shadows, provided a delightfully different reflection. The mirror was often in a gilded frame with an eagle on the cresting, an affectation the English took from the French Empire under Napoleon.

The eighteenth century was the golden age of the decorative mirror. Through the creative efforts of such famous designers as Thomas Chippendale, George Hepplewhite, Thomas Sheraton, and Robert Adams, mirrors brought a new dimension to the home. With the availability of clear, bright, distortion-free looking glasses, the homeowner, at least of the upper middle class, discovered that the new elements of brilliancy and variety of reflections had some advantages over commoner kinds of wall decoration. Properly placed in a room, a mirror was often superior to a picture. Mirrors filled the walls of stately halls with the light they gathered from windows at daytime and from sparkling chandeliers at night. They replaced the heavy woven tapestries of the past and competed successfully with painted vistas of landscapes or cities. The room opened to a wider world, and within the security of the home the illusion of being outside could be enjoyed.

The Seventeenth and Eighteenth Centuries

Western European society in the eighteenth century embraced the physical properties of the mirror and abandoned interest in its metaphysical characteristics. The mirror no longer possessed transcendental traits; it had been secularized. Beyond its aesthetic contribution, society sought the mirror for its ability to create an image that corresponded accurately to what others saw. Society was interested in perceiving the objective attributes of its identity. The new mirror taught self-observation.

The mirror of the eighteenth century was also turned toward the heavens, and, as the eye of the reflecting telescope, it initiated the exploration of other worlds. On March 13, 1781, the planet Uranus was revealed by the light-gathering power of the mirror. Sir William Herschel, the famed astronomer, had constructed a reflecting telescope containing a parabolic mirror about six and one-half inches in diameter which made the discovery. It was the first planet to be discovered in modern times. The known planets to that day (Mercury, Venus, Mars, Jupiter and Saturn) are visible to the naked eye, and they had been observed and studied by the ancient world. Uranus can only be seen through a telescope.

Quite suddenly, near the close of the eighteenth century, the mirror revealed another personality, a scientific personality that would make a vital contribution to our understanding of the universe. The scientific capabilities of the mirror did not suddenly appear overnight. They had developed slowly over two millennia to contribute finally to the means for a giant step in man's comprehension and exploitation of nature, not only with the reflecting telescope but also with other techniques that matured during the nineteenth and twentieth centuries.

NOTES AND SOURCES

1. Charles W. Cole, *Colbert and a Century of French Mercantilism*, 2 vols. (Hamden, Conn.: Archon Books, 1964), 2:304–14.
2. Ibid., p. 318
3. Johann Beckmann, *A History of Inventions, Discoveries and*

Origins, 2 vols., tr. William Johnston (Amsterdam: B. M. Israel, 1974), 2:80.

4. Quoted in R. W. Symonds, "Early English Mirrors," *Connoisseur* 96 (1935), 315.

5. Martha Fischer, "My Friend, the Connoisseur, Reflects on Mirrors," *House Beautiful* 62 (1927):592.

6. Quoted in R. W. Symonds, "English Looking Glasses," pt. 1, *Connoisseur* 125 (1950):9.

7. Samuel Pepys, *Diary*, 11 vols. ed. Robert Latham and William Matthews (Berkeley and Los Angeles: Univ. of California Press, 1970–83) 5:347.

8. Margaret McDonald-Taylor, "Mirrors of High Fashion," *Country Life* 136 (1964):970.

9. *Richard Steele's Periodical Journalism 1714–16*, ed. Rae Blanchard (London, Oxford Univ. Press, 1959), p. 121.

10. Quoted in Symonds, "Early English Mirrors," p. 317.

11. Margaret Jourdain, "Some Decorative Mirrors of the XVIIIth Century," *Apollo*, Aug. 1941, p. 51.

12. Ibid.

Other sources for this chapter include: Louise Ade Boger, *The Complete Guide to Furniture Styles* (New York: Scribners, 1959); William L. Bottomley, "Mirrors in Interior Architecture," *Architectural Forum* 57 (1932): 297–302; R. W. Symonds, "English Looking-Glasses," pt. 2, *Connoisseur* 125 (1950):82–87; Abraham Wolf, *A History of Science, Technology and Philosophy in the 18th Century* (New York: Macmillan, 1939).

PART
TWO

THE BURNING MIRROR

BEFORE DESCRIBING the events that led to Herschel's re-
flecting telescope and its growth and impact on the future of astronomy, we must return to the history of the
burning mirror. As early as 1000 B.C. the Chinese and the
Olmec Indians used mirrors to start fires. The ancient
Greeks were also well aware of this phenomenon. Indeed,
Archimedes is credited with using mirrors to set fire to the
Roman ships in the harbor during the siege of Syracuse by
Marcellus. Could this feat have been carried out? This question has stirred controversy among scientists since the seventeenth century, and it is still debated. The interest it created
and the experiments it generated on how mirrors may concentrate sunlight and cause burning at a distance have led to
the solar engines, cookers, and furnaces developed over the
past 150 years to utilize the sun's energy. Furthermore, the
idea of a concave mirror as the eye of the telescope, that is,
the reflecting telescope, arose from the investigation of the
feasibility of Archimedes' feat.

Archimedes' exploit was not documented until three
centuries after it occurred when Lucien, speaking of Archimedes, said that at the siege of Syracuse, he reduced the
Roman ships to ashes by a singular contrivance.[1] Galen also
noted that Archimedes fired the ships of the enemies of Syracuse with burning mirrors.[2] The first quantitative reference
came during the sixth century A.D. when Anthemius of

The Mirror and Man

Tralles asked, "How shall we cause burning by means of the sun's rays in a given position, which is not less distant than the range of bowshot?"[3] This was the assumed distance of the ships from shore. Anthemius pointed out that Archimedes could never have set fire to the Roman ships in the harbor with a single concave mirror because its focal length would have had to be equal to bowshot range and the area of the reflector gigantic, a condition beyond the resources of that time. Anthemius suggested that the feat may have been accomplished by a sufficient number of small plane mirrors arranged to reflect the sun's rays to a single point that, under favorable conditions, could ignite the wooden side of a ship.[4] The inclination of these mirrors would approximate a spherical or parabolic curved surface.

The subject remained relatively dormant until the twelfth century, when it was revived by Eustatheus, Zonares, and Tzetze. Of the three, Tzetze provided the most detailed statement which he extracted from the work of Pappus, a third-century Greek mathematician. Tzetze wrote, "When Marcellus had placed the ships a bowshot off, the old man (Archimedes) contrived a hexagonal mirror. He placed at proper distances from the mirror, smaller mirrors of the same kind, which were moved by means of their hinges and certain square plates of metal. He placed it in the midst of the solar rays. . . . The rays of the sun being reflected by this, a dreadful fire was excited on the ships, which reduced them to ashes at the distance of a bowshot."[5]

It should be noted that the sun is imaged as a circular disk at the focus of a parabolic mirror, rather than a point, whose diameter increases as the focal length of the mirror increases. This results in the dissipation of the concentrated heat over the large disk. Actually, the diameter of the disk can never subtend an angle of less than 32 minutes (about one half a degree) from the center of the mirror. This analysis was developed by Descartes in his *Dioptrique*. He concluded that a mirror with a very long focal length could not heat any more strongly than the direct rays of the sun because of the large size of the focused spot. Thus he rejected

The Burning Mirror

the possibility of Archimedes' feat.[6] But Descartes blundered in one essential aspect of his analysis when he stated that the intensity of the focal spot is independent of the absolute size of the mirror, so that a larger mirror (of the same focal length) will burn no more violently than a very little mirror. Descartes was unaware that as early as the second century B.C. Diocles had demonstrated that as the surface area of the mirror is increased (i.e., its aperture), the more sunlight it gathers, producing a more intense focal spot. So great was Descartes's fame that Archimedes' feat was treated as a fable for many years.

A contemporary of Descartes, Father Athanasius Kircher, whose work went unnoticed at that time, contradicted the former's analysis by the following experiment. He reflected sunlight progressively from one through five mirrors, each about one-foot square, on the same area at a distance of more than one hundred feet and noted that the heat from one mirror was somewhat like direct sunlight; from two, a little warmer and so forth until from five mirrors the heat could not be tolerated at all. He concluded that if one wanted to burn something at a distance, a large number of plane mirrors, properly arranged, could be better than almost any curved mirror.[7] So convinced was Kircher of the practicability of using a combination of plane mirrors for combustion that he suggested future mathematicians look further into this matter.

The man who undertook the challenge was Georges-Louis Leclerc, comte de Buffon, whose experiments in 1747 confirmed the possibility of Archimedes' feat.[8] Buffon, director of the Jardin des Plantes in Paris, was a practical scientist who believed in experimental methods.

He attached ordinary looking glasses, eight by six inches, to a frame; and each mirror, as well as the supporting frame, was capable of motion in every direction. With 40 of these mirrors, carefully tilted, he set fire to tarred beachwood (to simulate the side of the ancient ship) at a distance of 66 feet. Using 98 mirrors, he ignited a plank smeared with tar and brimstone at 126 feet. Increasing the number to 128, he ignited a plank of tarred fir at 150 feet. It blazed up at once over a space sixteen inches in diameter. After his experiments, Buffon noted: "It is certain that Archimedes was able to do with metal mirrors what I have done with glass ones; it is certain that his understanding was more than enough for him to imagine the theory that guided me, and the mechanism that I have made, and that therefore we cannot refuse him the title of the first inventor of these mirrors."[9] One hundred and twenty years after Buffon, John Scott of Scotland confirmed the former's results with a similar experiment.[10]

Interest in the subject continues to the present day, but experiments over the last twenty years have not added anything to these earlier results. It is likely that Archimedes used burning mirrors only once as a weapon of war against the Romans. The Roman sailors may have assumed that an attempt was being made to blind them, for they would not be aware of the burning capability at so great a distance. The boats probably ignited and the fire went unnoticed by the crews until it was too late. Panic would have reigned and the fleet could easily have succumbed.

The controversy surrounding Archimedes' feat that arose during the seventeenth century sowed the seeds for practical solar energy applications by using mirrors to collect and concentrate the sun's rays to produce high temperatures. Solar-driven engines, cooking by sunlight, solar furnaces, and power generation were made possible. Toward the close of the century when Father Kircher performed his mirror experiments, an optician named Villete from Lyons constructed iron parabolic mirrors as solar energy concentrators that he used to smelt iron, copper, and gold. His reputation spread, and his mirrors were used for this purpose not only in France but in other countries of Europe and Asia. One of his mirrors, still extant, is about thirty-two inches in diameter with a focal length of about thirty-eight inches. Another Frenchman, Dufay, also experimented with smelting metals by concentrating solar energy. In the eighteenth century Buffon used the Villete mirror, which was then at the

Academie, to gain preliminary information for the design of his bank of mirrors. The French started a "burning mirror" tradition for solar energy utilization that is with us today.

During the seventeenth and eighteenth centuries the burning mirror appears to have been as much a novelty as a practical device. It was not until the middle of the nineteenth century that the mirror became a key accessory to a practical device: the solar engine. Its inventor was August Mouchet, who started his solar work in 1860. He took out his first patent on a solar engine in 1861; it was a device to produce steam for mechanical energy.[11] Basically, it comprised a boiler (three and a half pint capacity) placed at the focus of a curved reflector. The boiler consisted of two cylindrical concentric vessels with domed tops and the water space between them. The outer vessel was sixteen inches high and centered at the focus of the reflector. With the mirror oriented toward the sun, the water boiled in one hour from an initial temperature of fifty degrees Fahrenheit.

In August 1866 Emperor Napoleon III saw Mouchet's first solar engine at work in Paris. Greatly impressed, the emperor provided Mouchet with government funds to construct another sun boiler, and this was completed in 1872. It was a superb achievement that attracted widespread interest for many years. "The traveler who visits the library of Tours sees in the courtyard . . . a strange looking apparatus. Imagine an immense truncated cone, a mammoth lampshade, with its concavity directed skyward. This apparatus [mirror] is of copper, coated on the inside with very thin silver leaf. On the small base of the truncated cone rests a copper cylinder [with hemispherical cap], blackened on the outside, its vertical axis being identical with that of the cone . . . and is covered with a bell glass of the same shape." This is the water boiler which, heated by the sun, generates steam for power. "The reflector, which is the main portion of the generator has a diameter of 2.6 meters at its large and one meter at its small base, and is 80 cm. in height, giving 4 square meters of reflecting surface."[12]

A conical mirror, it focuses sunlight along a line coinci-

The Mirror and Man

dent with its vertical axis, thus heating the cylindrical boiler along its length to provide maximum intensity. On a sunny day it was able to bring to a boil, in forty minutes, about twenty-five quarts of water from room temperature to produce steam under high pressure. It rated about one-quarter horsepower. Less successful was the work of Carl Guntner of Austria, who developed a plane mirror system which followed the sun to keep constant reflection on the surface of a boiler. He started his experiments in 1854, and his apparatus was exhibited at the 1873 Vienna Exposition.

Mouchet continued to build larger systems, and his work started a worldwide trend. In the United States, John Ericsson, a versatile engineer and successful inventor, built a solar engine in 1872 using a large spherical reflector that concentrated solar radiation onto one end of a cylindrical boiler. In India the Englishman W. Adams, who was deputy registrar of the High Court in Bombay, built several solar engines in the latter 1870s. A great proponent for the use of solar energy in underdeveloped countries, he suggested many uses for solar heat, and he built practical solar cookers.[13] Solar cookers are still used in developing countries, especially where wood is scarce. One such region is the province of Gansu in China, where the Hui, a Muslim minority, use such a device (fig. 23). A tilted solar mirror focuses the sun's rays on a teapot in which water can be brought to a boil in twenty minutes.

A unique application of the sun-powered engine was devised by Abel Pifre of France. Using a parabolic reflector about eleven feet in diameter, he collected sufficient energy to run his one-horsepower printing press. A significant improvement in the solar-powered steam engine was made by an English inventor, Aubrey G. Eneas, who lived in Boston. It consisted of a gigantic inverted cone lined with thousands of flat glass mirrors that concentrated sunlight upon a boiler at the center of the reflector. It was made to follow the sun across the sky by a modified clockwork mechanism. The first of these machines was installed in South Pasadena, Califor-

The Burning Mirror

Solar cooker. (Photo courtesy of Herman How-Man Wong)

nia, in 1901. Another was installed in Tempe, Arizona (fig. 24). When used on a sunny day, this machine could pump enough water to irrigate eighty acres of dry but fertile Arizona land. Soon other machines appeared in arid areas of the southwest United States, and they were used extensively until about 1906, when the introduction of small gasoline engines and electric motors made irrigation possible at lower cost and in all kinds of weather.

If solar engines soon disappeared in most highly industrialized nations, they continued to play a role in the desert regions of underdeveloped countries and even in underdeveloped areas of some industrial nations. Solar engines were constructed in Australia and Mexico. In 1913 the engineer Frank Shuman built a fifty-five-horsepower solar irrigation pump near Cairo, Egypt. This was a giant step, for it meant construction of much larger, more efficient mirror areas. Mouchet's reflector of four square meters had produced only about one-quarter horsepower.[14]

Solar water pumps continued to be used in the desert

Solar-powered steam engine. (Courtesy of Arizona State University)

regions of the world, and in the 1940s the USSR built several sophisticated solar stills in its desert areas. These were used, among other things, for the production of ice by absorption refrigeration and the distillation of salt water at the rate of about one ton of pure water during a ten-hour sunny day. Concave mirrors were used to focus sunlight on the boilers to generate steam. The largest was a reinforced concrete cup 33 feet in diameter lined with silvered glass mirrors, providing a reflecting surface of 800 square feet. The USSR is still employing solar stills in Soviet Central Asia and the Transcaucasus. The solar still on the Shafakan State Farm produces four metric tons of fresh water a day at only two rubles per cubic meter as opposed to 100 rubles per cubic meter for water obtained by other means.

In France, Felix Trombe, after many years of effort,

The Burning Mirror

completed an immense solar furnace in 1970. It is used for melting refractory material. Located in Odeillo, the parabolic mirror of the furnace, built into a mountainside, is about 22,000 sq. ft. It is capable of producing a temperature in excess of 7,000 degrees Fahrenheit at the focus.

With the widespread consumption of oil, its attendant shortage, ever-increasing price, and the uncertainty of the future of nuclear energy, a number of electrical utilities in the American Southwest and the U.S. Government began to turn to solar energy in the early 1970s. The idea of myriads of mirrors used in a gigantic version of the Archimedian approach demonstrated by Buffon became part of a system known as the "power tower."

It started when the Department of Energy accepted the idea of a New Mexico electric utility company to provide a booster system for existing power plants that would utilize acres of plane mirrors reflecting the sun's heat to a boiler full of water on top of a 1,000-foot tower. The resulting generated steam would drive electricity producing turbines. The company hoped to combine this system with standard power plants fueled by natural gas or oil and thus reduce consumer costs and fuel use when the sun shone. They proposed a demonstration model that would use 5,000 plane mirrors, heliostatically driven, to track the sun and reflect its rays to the power tower situated in the center. The mirrors and the generating plant would cover about two hundred acres of land. A survey by the company indicated that some six hundred plants in the Southwest alone could be adopted to what they called the "solar hybrid repowering system." An enormous amount of heat would need to be generated at the power tower by the mirrors to raise the temperature of the water to the 1,000 degrees Fahrenheit required by the steam-cycle power plant generators.

In 1978 the Department of Energy chose designs for a group of 735-feet-high power towers. These would rise from mile-square fields studded with 23,000 solar heliostat units. Each heliostat would carry an array of mirrors, 20 by 20 feet. Each field would generate 100 million watts (100 megawatts)

Power tower in a field of mirrors, or "heliostats." (U.S. Department of Energy and Martin Marietta Corp.)

of electricity (fig. 25). Before proceeding with this grand undertaking, an evaluation model was started in January 1979. The 10-Megawatt Solar Thermal Central Receiver Pilot Plant is located near Barstow, California, on 130 acres. It consists of 1,800 computer-controlled heliostats, each with 400 square feet of mirror surface, surrounding a 330-foot tower. Enough electricity is produced to meet the needs of a community of 10,000 residences. It cost over $130 million and has been in operation since 1982. This plant should provide impetus for future commercial applications in the 50 to 150 megawatt range.

NOTES AND SOURCES

1. Louis Dutens, *An Inquiry into the Origin of the Discoveries Attributed to the Moderns* (London: Printed for W. Griffin, 1769), p. 331.

2. Ibid.

3. George L. Huxley, *Anthemius of Tralles: A Study in Later Greek Geometry, Greek, Roman, and Byzantine Studies*, no. 1, 1959, p. 12.

4. Ibid.

5. John Scott, "On the Burning Mirrors of Archimedes," *Transactions of the Edinburgh Royal Society* 25 (1867–69):123.

6. W. E. K. Middleton, "Archimedes, Kircher, Buffon, and the Burning Mirrors," *Isis* 52 (1961):534.

7. Ibid., p. 535.

8. Ibid., p. 537.

9. Ibid., p. 540.

10. Scott, "Burning Mirrors," p. 147.

11. A. S. Ackermann, "Utilization of Solar Energy," *Annual Report*, Smithsonian Institution, 1915, p. 149.

12. Ibid., pp. 149, 150.

13. Ibid., p. 153.

14. Dennis Hayes, "Solar Power in the Middle East," *Science* 188 (1975):1261.

Other sources for this chapter include: W. Adams, "Cooking by Solar Heat," *Scientific American*, June 15, 1878, p. 376; Thomas W. Africa, "Archimedes through the Looking-Glass," *Classical World* 68 (1975):305–8; Dan Behrman, "Solar Energy Claims a New Place in the Sun," *UNESCO Courier*, Jan 1975, p. 24; Guy Benveniste, "Burning Glasses: From Archimedes to Lavoisier," *The Sun At Work* 1 (June 1956):4–6; A. C. Claus, "On Archimedes' Burning Glass," *Applied Optics*, Oct. 1973, p. A14; Willi M. Conn, "Recent Progress in Solar Furnaces for High Temperature Research and Development Work," *Journal of the Franklin Institute* 257 (1954):1–11; Diocles, *On Burning Mirrors*, tr. G. J. Toomer (New York: Springer-Verlag, 1976); J. Harris and Desaguliers, "An Account of Experiments Made with Mons. Villette's Burning Concave," *Philosophical Transactions* 30 (1719):967–77; Klaus D. Mielenz, "Eureka!" *Applied Optics*, Feb. 1974, p. A14; Vadim Orlov, "Hunting the Sun in the Steppes of Central Asia," *UNESCO Courier*, Jan. 1974, pp. 33–36; D. L. Simms, "Archimedes and the Burning Mirrors of Syracuse," *Tech-*

nology and Culture 18 (1977):1–24; O. N. Stavroudis, "Comments on: On Archimedes' Burning Glass," *Applied Optics*, Oct. 1973, p. A16; H. Tabor, "A Solar Cooker for Developing Countries," *Solar Energy* 10 (1966):153–57; M. Telkes, "Solar Cooking Ovens," *Solar Energy* 10 (1966):2–11; Felix Trombe, "Solar Furnaces and Their Applications," *Journal of Solar Energy & Engineering* 1 (1957):9–15; Felix Trombe and Albert Vinh, "1000 KW Solar Furnace in Odeillo," *Solar Energy* 15 (1973):57–61; How-Man Wong, "Peoples of China's Far Provinces," *National Geographic* 165 (1984):-283–333; John Yellott, "Solar Energy in Arizona," *Arizona Highways*, Aug. 1975, pp. 6–14.

THE REFLECTING
TELESCOPE

THE EARLIEST TELESCOPE was the refracting telescope, one that uses lenses only. It was invented by Hans Lippershey in Holland in 1608. The following year Galileo, hearing of Lippershey's invention, made his first telescope and used it for astronomy. He fitted one end of a tube of lead with a convex lens, the objective, and the other end with a concave lens, the eyepiece. The refracting telescope, however, had serious shortcomings. The enlarged image had poor definition and it was surrounded by a ring of colored fringes. The reflecting telescope, which uses a curved mirror for the objective, offered a way to eliminating these and other deficiencies.

The early seventeenth-century controversy regarding Archimedes' feat also resulted in the idea of the reflecting telescope, the searching eye of the universe. While Father Kircher and Descartes were disagreeing about the possibility of Archimedes having burned the Roman fleet, Francesca Bonaventura Cavalieri, a disciple of Galileo and professor of mathematics at Bologna, was also studying the properties of mirrors to determine the validity of the feat. He tried various curved mirror designs and mirror combinations to see how Archimedes may have transferred the focused hot spot of the sun from the vicinity of a parabolic mirror to a distance of several hundred feet without loss of intensity. This led to a design for a reflecting telescope. His results were published in his book *Specchio Usterio* (*Burning Mirror*), in 1632.[1]

He restated in a comprehensive manner the reflecting properties of mirrors generated from the parabola, ellipse, and hyperbola, from both their concave and convex sides. Then he was able to show what combination of reflecting surfaces are required for transforming rays that are parallel, convergent, or divergent into rays that are convergent, divergent, or parallel. This was of fundamental importance because it introduced the concept of employing a train of mirror combinations to transfer light rays from A to B in a most efficient manner, providing a converging, diverging, or parallel beam at B as required for a particular application. This idea was essential to the design of future reflecting telescopes, including those that use perforated mirrors.

Figure 26a shows a concave parabolic mirror looking at the sun along the axis XM and focusing the reflected rays at I. In accordance with his description, Cavalieri placed a small convex parabolic mirror with its focal point coincident with I. The converging reflected rays from the main mirror impinge on the convex surface of the secondary mirror and are reflected parallel to the axis of the latter toward P, where they start the desired fire. Figure 26b shows a perforated primary mirror and a convex parabolic secondary mirror whose focal point is coincident with the main mirror, which is behind the opening. Rays from the sun reflected from the large mirror converging through its hole toward the focus are rereflected from the small mirror toward the target P. Cavalieri demonstrated a number of these mirror arrangements.

Toward the end of his discourse on burning mirrors, he addressed the idea of the reflecting telescope: "I could also say that the effect of the telescope could possibly be obtained from a combination of these mirrors."[2]

Cavalieri considered making a simple telescope by using a concave mirror to look at a desired scene and viewing the resultant image through an ocular lens, or eyepiece, that would act as a magnifier. (The magnifying power of a telescope is obtained by dividing the focal length of the primary

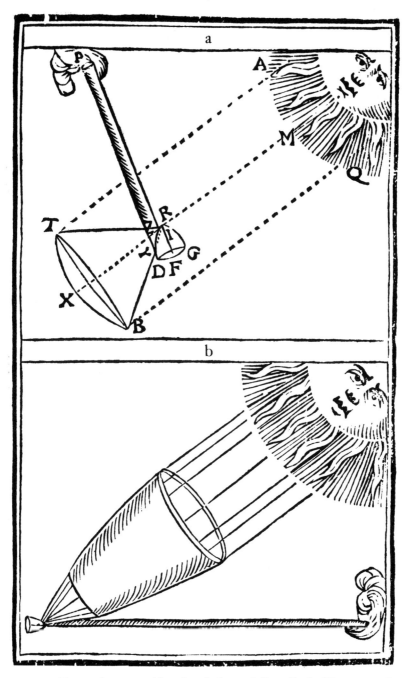

Two mirror combination designs of Cavalieri. (Courtesy of the History of Science Collections, Cornell University Libraries)

mirror by the focal length of the ocular, which is very short, less than half an inch.) Cavalieri anticipated the Newtonian telescope design, one of four basic types further described below, when he suggested that the ocular be placed at right angles to the axis of the concave mirror and the incoming image viewed via a forty-five-degree mirror placed in its path. The ocular and the inclined mirror could be moved as a unit to focus the image (fig. 27).

Despite his basic contribution to the design of the reflecting telescope, Cavalieri felt the idea whimsical: "to give satisfaction . . . to these frivolous people who crave cake instead of bread. For in my view [the reflecting telescope] will never match the excellence of the refracting telescope either by combination of mirrors or by the addition of lenses as anyone who wishes to try will, I believe, find out."[3]

A contemporary of Cavalieri, Marin Mersenne of France, who was also examining the problem of the burning mirror of Archimedes, was familiar with the former's work and became interested in the reflecting telescope. Mersenne was a scientist and philosopher of note. He presented his contributions to the development of the telescope in his *Harmonie Universelle* in 1630. These consist mainly of two designs that he initially considered for the Archimedian problem.

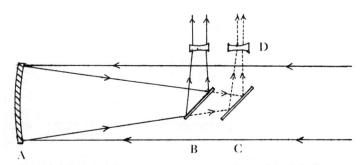

Cavalieri's "Newtonian" reflector. Primary concave mirror A converges incoming light toward inclined flat mirror that may be moved with eyepiece D to a position (e.g., B or C) to provide focused image to the eye.

One (fig. 28a) shows a concave parabolic mirror *ESTL* which has a hole at *ST* and is cofocal with a smaller concave parabolic mirror *ACP*. Light rays, parallel to the axis, striking the concavity of the primary mirror, say at *E*, are reflected through the common focus to point *P* of the secondary and then reflected again as *PN* parallel to the axis of the system and through the opening *ST* to the target. Such a system, Marsenne noted, can be used not only for lighting fires at a distance, but "it will be more to the point to use such an invention in making telescopes."[4]

Mersenne's second design substitutes a convex secondary mirror for the concave one (fig. 28b). These two designs

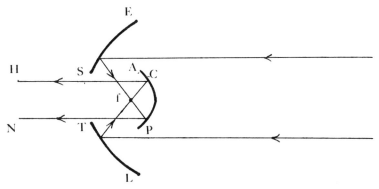

Mersenne's reflector combination designs: concave-concave parabolic combination.

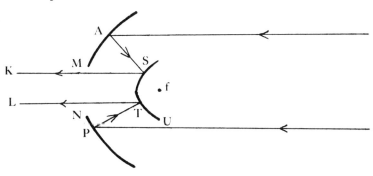

Mersenne's reflector combination designs: concave-convex parabolic combination.

are prototypes of the future Gregorian and Cassegrain tele-scopes (see Fig. 29a,c). Readily acknowledging the priority and influence of Cavalieri's efforts, Mersenne's contribution lies in the wide dissemination of his two designs by their repeated publication. He did add some original ideas essen-tial to telescope design, conceiving a rigid frame to hold the mirrors together and keep them at their proper distance and suggesting that this structure be enclosed in a tube blackened on the inside to prevent scattered light from getting into the eye. (Except for the mention of this historical point, it is not the intent of this book to go beyond the evolution of the mirror as the eye of the telescope. Other components in the making of the modern telescope are equally important.)

Mersenne, like Cavalieri, did not believe the reflecting telescope would provide an image equal to that produced by the refracting telescope. The problem was the great difficulty in accurately grinding and polishing not only the parabolic, elliptical, or hyperbolic mirror but the less difficult spherical surface. Indeed, Cavalieri first tried to make a parabolic mir-ror himself and later asked an optician to make one, but both failed. There is no evidence that Mersenne ever made a mir-ror to check his designs. Despite skepticism toward their own designs, Cavalieri and Mersenne laid out the essentials of the reflecting telescope in three basic designs before the men who were later to invent them.

In 1663 a Scottish mathematician, James Gregory, pro-posed the construction of a reflecting telescope in his *Optica Promota*. Five years later he made this instrument by placing a concave metal mirror with a hole in the center at the end of a short, wide-mouthed tube. Opposite the hole he positioned a very small concave mirror so that the light rays received by the primary mirror were reflected to the little mirror and sent from there to the eyepiece through the hole of the pri-mary mirror (fig. 29a). This small telescope magnified as well as a refracting telescope twice its length, and it had the ad-vantage of compactness. It could be held in the hand and did not require an elaborate mount.

At about the same time Isaac Newton, who was study-

The Reflecting Telescope

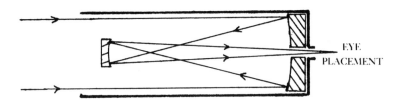

EYE
PLACEMENT

The Gregorian telescope.

EYE PLACEMENT

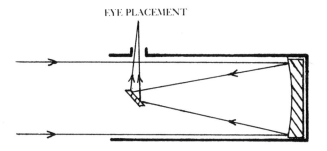

The Newtonian telescope.

ing the nature of light, came to the conclusion that the color fringes produced by the dispersion of white light through a lens (chromatic aberration) were incurable and would forever limit the image definition of a refracting telescope. After reading about Gregory's design, he believed that a reflecting telescope was the only way to eliminate this problem. He set out to build one of his own design which he completed in 1668, the same year Gregory finished his model. Newton fabricated a small spherical mirror of speculum metal whose focal length was 6⅓ inches and aperture 1⅓ inches which he mounted on one end of a 6-inch tube. Within the tube and close to the open end he suspended a very small inclined plane-mirror that reflected the converging rays from the spherical mirror at right angles through an opening in the side wall of the tube where the ocular was placed (fig. 29b). Since the ocular had a focal length of ⅙ of an inch, the magnification of this miniature telescope was 38. (The magnification of a telescope is f.l. of mirror / f.l. of ocular, or 6 ⅓ divided by ⅙.) When he viewed the crown on a weathercock about 300 feet away with his 6-inch-long reflecting telescope, Newton noted that the diameter of the image was 2½ times greater than that provided by a refracting telescope about 2 feet long.[5] He also reported that he was able to read a page of the *Philosophical Transactions* at a distance greater than 100 feet.[6] Later, Newton built a second telescope with a two-inch-diameter spherical mirror.

The Gregory telescope became the plaything of the eighteenth century, but Newton's design, after some neglect, was adopted for astronomical telescopes because looking down into the eyepiece instead of looking up was an easier way to view the stars. The Gregory design, more suitable for terrestrial use, was quite popular. These telescopes were exhibited in fine shops as articles of luxury, and the rich bought them as novelties. Gregorian telescopes in stands were regular display items in instrument makers' shops. They continued to be made in England and elsewhere until the first quarter of the nineteenth century. Even very tiny telescopes were fabricated. One of these is in the Fuggers Collection at Augsburg. It has a six-inch focal length and a one-inch aperture and is hidden in the handle of a walking stick.[7]

A third kind of telescope was constructed in 1672 by Guillaume Cassegrain. This telescope was about the same as Gregory's except that a small convex mirror replaced the concave one (fig. 29c). This brought the focused image to a point much farther behind the primary mirror, providing more flexibility for the eyepiece placement. Cassegrain later modified his instrument by incorporating Newton's idea of a small inclined plane-mirror after the convex secondary to redirect the focused image through the side of the tube. This eliminates the need to perforate the primary mirror. The Cassegrain design, like Newton's, survived to be used by modern astronomers. Modern versions of the Gregorian and Cassegrain telescopes use concave elliptical and convex hyperbolic secondary reflectors.

These three inventors, however, could not grind para-

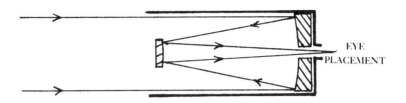

The Cassegrainian telescope.

bolic reflectors. They were obliged to use spherical mirrors for their primary reflectors. To minimize the inherent spherical aberration that produced somewhat blurry images, they made the mirror apertures quite small. Indeed, the technique for grinding and polishing sufficiently accurate parabolic mirrors with sufficiently large apertures had to be developed before the reflecting telescope could be called superior to the refractor. This happened in 1721 when the scientist John Hadley succeeded in grinding a parabolic mirror for a Gregorian telescope he had constructed. It was the first reflecting telescope that could compete with refracting ones.[8] His parabolic mirror was 6 inches in diameter and had a focal length of 5¼ feet. When he demonstrated it to the Royal Society of London, it compared favorably with the refracting telescope they owned, which was a little wider, but it was 123 feet long while Hadley's was only 6 feet long. James Pound, an astronomer who made several observations with Hadley's reflecting telescope, uttered a prophetic statement in the *Philosophical Transactions* of 1723. After praising Hadley's telescope, he noted "that the old [refracting] telescope will be for the most part laid by, and this [reflecting] one will be chiefly in use among practical astronomers."[9]

It was the famed astronomer Sir William Herschel who finally demonstrated the power of the reflecting telescope as a searcher of the heavens. Herschel recognized the relationship between the aperture of a telescope and its light-gathering power, the key consideration for penetration deep into space. For example, if you triple the diameter of the primary

The Mirror and Man

reflector, the image will be nine times brighter. He was the first man to write about this relationship in 1789 in his *On the Power of Penetrating in Space by Telescopes.*[10]

An early interest in astronomy led him to search for a suitable telescope. After many trials with the different kinds of reflecting telescopes, he decided in favor of the Newtonian one and become an expert in its construction. He made hundreds of mirrors for telescopes, some of them up to 19 inches in diameter. As his fame spread, he sold Newtonian reflectors of 7-foot focal length with 6½-inch aperture up to 10-foot focal length and with 8⅘-inch aperture to buyers throughout Europe.

Herschel himself, using a 7-foot focal length telescope with a 6¼-inch aperture, made an astounding discovery. On March 13, 1781, he noticed in the constellation Taurus a "nebulous star or perhaps a comet" appearing as a distinct disk. It proved to be a planet beyond Saturn. This planet, Uranus, the first to be discovered in modern times, was revealed by the light-gathering power of a mirror.[11]

The impression made in scientific circles by Herschel's discovery was extraordinary. Demand for reflecting telescopes grew as astronomers recognized the confines of the refracting telescope which could not then be made in excess of five inches in diameter, thus limiting its light-gathering capabilities. George III of England ordered five reflecting telescopes, and Herschel made seventy-six for many dignitaries, including the king of Spain, the emperor of Austria, and Catherine the Great, of Russia.

Now a wealthy man, Herschel settled at Slough, where he constructed his largest telescope. It employed a giant metal reflector of 40-foot focal length encased in a 40-foot cylinder, 54 inches in diameter. The design became known as Herschelian (fig. 29d). He used one parabolic mirror in a slightly inclined position. "It consists," he wrote, "in looking with the eye-glass, placed a little out of the axis, directly in at the front, without the interposition of a small speculum; and has the capital advantage of giving us almost double the light of the former constructions."[12] Because the observation

The Reflecting Telescope

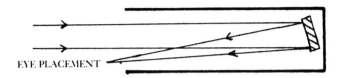

The Herschelian telescope.

position is awkward and the observer's head obstructs some of the light, this design is not much used today.

Herschel's mirrors opened the skies and laid the foundation for modern astrophysics. Not only did he discover Uranus and two of its satellites, but he also discovered and cataloged the nebulae and a large number of double stars and star clusters.

Herschel's successes motivated lens makers to try to make larger-diameter lenses to compete with reflecting telescopes, but they reached their limit with the construction of a refractor with a 40-inch diameter objective lens. The construction of a 72-inch diameter primary mirror by Lord Rosse, William Parsons, in 1845 for his reflecting telescope left the refractor as second best for astronomical work. This 72-inch-diameter telescope with a tube 50 feet long became known as the Leviathan of Parsonstown. The mirror was the biggest ever cast of metal (speculum) and it weighed 4 tons. It was the last of the great metal mirrors. With it, Rosse contributed to the knowledge of the heavens by distinguishing between various kinds of nebulae and galaxies. He revealed for the first time the characteristic spiral shape of many of them.

By this time the silvering of glass mirrors had replaced the tin amalgam process. In 1835 Justus von Liebig of Germany discovered the process of silver deposition on glass to produce mirrors of superior brilliance. This did away with the poisonous mercury required in the earlier process. Thomas Drayton of England initiated the first production of silver-backed mirrors in 1843. Metal mirrors for telescopic use gave way to silvered glass mirrors.

The Mirror and Man

Toward the end of the nineteenth century, with the mysteries of the universe unfolding, astronomers extended the goals of their research beyond the brighter stars and the possibility of locating galaxies unseen till then. Their reflecting telescopes had viewed to their limit without disclosing the structure of the galaxy, yet what they saw did hint at the vastness of the universe. This stimulated the need for larger mirrors in telescopes because on size alone depends not only the amount of light received from a star but the ability to resolve two stars very close together. As noted earlier, the practical upper limit for the diameter of the lens was reached with an aperture of 40 inches. There are two reasons for this. Because lenses can only be supported around their outer rims, they tend to sag under their own weight. Their optimum performance deteriorates as the weight increases with the increased aperture. The lens, through which the light passes, must also be free of bubbles and striations. And the larger the lens, the thicker the glass. This leads to loss of light because of absorption.

Mirrors, on the other hand, which are made of glass suitably silvered on the front surface, do not have to be concerned with the internal uniformity or thickness of the glass. In addition, a mirror can be supported over the entire back surface to eliminate the sag problem. It is important to point out, however, that to obtain the same optical performance from a mirror as from a lens, the reflecting surface must be four times as accurately ground as the surface of a refracting lens. Even Newton ground his tiny reflector to within one-thousandth of an inch of the true spherical curve. The curve of the great 200-inch parabolic mirror of the Mt. Palomar telescope has an error of less than two-millionths of an inch. The reflector enjoys another important advantage. Although a refractor can be used for visual observation, it is not adopted for photographic work, which is an important aspect of astronomical observation. The reflector is at its best in photographic applications.

The first giant telescope of the twentieth century was designed and constructed by George Ellery Hale. This

The Reflecting Telescope

200-inch-diameter reflecting telescope took nineteen years to complete. Dedicated in 1948, after Hale's death, it represented the greatest accomplishment of his life. Previously he had completed the 40-inch refractor at the Yerkes Observatory and the 60-inch and 100-inch reflecting telescopes at Mt. Wilson, California.

The 200-inch parabolic mirror, 25 inches thick and weighing 30 tons, is made of Pyrex with a honeycombed back to minimize its weight. It has a focal length of 55 feet. The design is Cassegrainian combined with the Newtonian viewing mode with the attendant advantages of a short length in comparison to its focal length and a reduction of weight in the tube structure.

The astronomical advances made possible by Hale's 100-and 200-inch reflecting telescopes were nearly all obtained by photographic means. This technique has an enormous advantage over the eye, not only in recording what the telescope sees, but by the exposure of film for long periods of time that can integrate the faint light of stellar objects to obtain well-defined pictures. The 100-inch telescope unveiled the true size and nature of our galaxy for the first time, and it revealed the scope of the universe beyond the galaxy. The motions of the outer galaxies presented the vision of an expanding universe. The Hale reflector made it possible to photograph stars that were only 30-billionths as bright as the faintest star visible to the naked eye. It detected faint galaxies hundreds of millions of light years away from us.

But this photographic astronomy has one handicap. The parabolic mirror proved extremely efficient so long as the observation was close to the optical axis of the mirror. Star images reflected farther and farther from the optical axis deteriorate from bright crisp points to blobs with taillike flares looking like commas, an aberration known as "coma." Also problems of astigmatism and curvature of the field further deteriorated the crispness and fidelity of the image. Thus, while a photographic plate may show tens of thousands of celestial bodies within the field of view of the telescope, only the relatively few stars near the center of the photographic plate will be sharp. Mapping the skies was inordinately difficult because only the center of the photographic plate could provide useful information.

Bernhard Schmidt, a German lens and mirror grinder, confronted these problems. Schmidt did not attempt to correct the image produced by the parabolic mirror. Instead, he used a spherical mirror because he realized that spherical aberration only will adversely effect its performance. To overcome this deficiency, he designed a thin transparent plate with a carefully calculated undulating surface. When this was placed perpendicular to the optical axis at the center of curvature of the mirror, it deflected incoming rays to a well-defined focus over a wide field. Interestingly, the photographic plate must be mounted over a curved surface to obtain aberration-free results.

The first Schmidt telescope camera was mounted at the Hamburg Observatory in 1931. Later it attained its full prominence when it became a valuable assistant to the 200-inch reflector at Mt. Palomar. Fondly called "the Big Schmidt," it was installed a quarter mile east of the great telescope in 1948. Its 72-inch spherical mirror and 48-inch correcting plate made possible a systematic photographic survey of the sky that would have taken perhaps a thousand times longer with the parabolic system. The largest Schmidt telescope now in operation has a 79-inch-diameter mirror and 53-inch corrector plate. It is located in the Karl Schwarzchild Observatory in Jena, East Germany.

Other important reflecting telescopic systems have been constructed during this century and contribute to our knowledge of the universe. Two accomplishments of the 1970s are particularly important: the colossal reflector telescope of the USSR and the Multiple Mirror Telescope of the United States. After sixteen years of construction, the Soviet Union put into operation in December 1975 the world's largest reflecting telescope atop Mt. Postukhov, in the Greater Caucasus. The parabolic mirror of this telescope measures 237 inches in diameter and it weighs 42 tons. This telescope can see 50 percent further into the universe than Mt. Palomar. It can, for example, see the light of a candle more than 15,000 miles away. The Russian astronomer Dr. E. Mustel suggests that the telescope will first be used to study the mysterious quasars, which emit enormous bursts of radio waves into deep space. After that, double stars that revolve around each other at great speeds will be photographed in detail, as well as stars and distant planets with powerful magnetic fields and unusual chemical substances.[13]

In the United States, astronomers have been studying various ways to build a telescope larger than that at Mt. Palomar. Early in the 1970s the telescope designers Aden Meinel and Fred Whipple and their associates discovered a new idea. Realizing the impracticality of fabricating a single mirror several times larger than George Hale's mirror (the goal was a 1,000-inch-diameter reflector), they hit upon the notion of assembling hundreds of parabolic mirrors, each about 72 inches in diameter, which would be equivalent in performance to one very large mirror when they acted together. They called this design the Multiple Mirror Telescope, or MMT.

A prototype of the MMT was started in 1972 under the direction of the University of Arizona and the Smithsonian Astrophysical Observatory and was dedicated on May 9, 1979. It is located at the Smithsonian's Mt. Hopkins station about forty miles south of Tucson, Arizona. The basic design is a cluster of six 72-inch Cassegrain telescopes with parabolic primary and hyperbolic secondary mirrors, arranged about a common axis (fig. 30) with the six reflected beams relayed to a single focal plane by auxiliary plane reflectors (fig. 31). A pyramid of mirrors at the center combines the beams and redirects the light, now a single image, to recording instruments located at the focal point behind the telescope. Accuracy of alignment is assured by an electronic control system that utilizes lasers and on-line computers. These superimpose images and then maintain them as the telescope changes its position to track stars. The combined light-collecting areas equal that of a 180-inch-diameter conventional telescope. The telescope, supported by an altitude-

Compound eye of the multiple mirror telescope.
(Photo courtesy of Gary Ladd)

azimuth mount, is enclosed in a rotating building. Even with the complicated system required to line up the MMT, its cost is about a third that of an equivalent 180-inch telescope. The total weight of the six mirrors is about four tons compared with about thirty-eight tons for a single solid blank of equivalent collecting area. The tube length of the MMT is shorter than an equivalent individual telescope, and it can be housed in a building considerably smaller than the conventional domed observatory.

The MMT is especially well suited for infrared observation; it is the largest telescope ever constructed specifically for this purpose. It will reveal new information about the birth of stars, and it will be able to better specify conditions

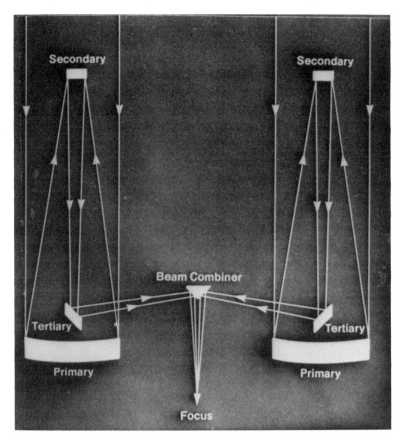

Schematic showing how light received by primary mirrors is brought to a focus. Light from a celestial object (white lines) is focused by each primary mirror so that it travels to secondary mirror, down to a third, then to a beam combiner. From there the six images come to a common focus.

at the heart of our own Milky Way and in cores of more active galaxies where enormous amounts of infrared radiation have been found. The study of cool stars, which are larger than our sun but about one-third its temperature, will be made possible, as will investigation of new stellar forms known as protostars, which can only be detected by their infrared radiation.

The first MMT, which marks the birth of a new generation of astronomical telescopes, is being followed by a larger design. A group at Kitt Peak National Observatory hopes to complete its work by the late 1980s on a new MMT design that will be equivalent to a single 600-inch-diameter mirror. The edge of the universe as we know it now will soon be enormously expanded.

NOTES AND SOURCES

1. Piero E. Ariotti, "B. Cavalieri, M. Mersenne, and the Reflecting Telescope," *Isis* 66 (1975):303–21.

2. Ibid., p. 314.

3. Ibid., p. 320.

4. Ibid., p. 318.

5. Bernard I. Cohen, *Isaac Newton's Papers and Letters on Natural Philosophy* (Cambridge: Harvard Univ. Press, 1958), p. 62.

6. Thomas H. Court and Moritz von Rohr, "A History of the Development of the Telescope from about 1675 to 1830 Based on Documents in the Court Collection," *Optical Society Transactions* (London) 30 (1928/29):219.

7. Ibid., p. 226.

8. Cohen, *Newton's Papers*, p. 41.

9. James Pound, "Concerning Observations Made with Mr. Hadley's Reflecting Telescope," *Philosophical Transactions* 32 (1723):382.

10. William Herschel, "On the Power of Penetrating in Space by Telescopes," *Philosophical Transactions of the Royal Society*, 1800, pp. 49–85.

11. Abraham Wolf, *A History of Science, Technology, and Philosophy in the 18th Century* (New York: Macmillan, 1939), p. 115.

12. Ibid., p. 116.

13. "New Russian Telescope," Sarasota *Herald Tribune*, Feb. 24, 1976.

Other sources for this chapter include: Phyllis Allen, "Problems Connected with the Development of the Telescope (1609–1687)," *Isis* 34 (1943):302–11; Larry Barr and Brian Mack, Chairmen/Editors, *Proceedings of the International Society of Optical Engineering*, vol. 444, *Advance Technology Optical Telescopes II*, Sept. 5–6, 1983, London; Nathaniel P. Carleton and William F. Hoffman, "The Multiple-Mirror Telescope," *Physics Today*, Sept. 1978, pp. 30–37; James Cornell, "The Multiple Mirror Telescope," *Smithsonian*, May 1979, pp. 42–50;

The Reflecting Telescope

D. L. Crawford, ed., *The Construction of Large Telescopes*, International Astronomical Symposium no. 27, Tucson, Ariz. (Apr. 5–12, 1965); G. Z. Dimitroff and J. G. Baker, *Telescopes and Accessories* (Philadelphia: Blakiston, 1945); Edward A. Fath, "The Reflector—the Telescope of the Future," *Scientific American*, July 12, 1913, p. 30; E. Finley Freundlich, "Development of the Astronomical Telescope," *Journal of Scientific Instruments Review* (London) 27 (1950):233–37; George Ellery Hale, "Building the 200-inch Telescope," *Harper's*, Nov. 1929, pp. 720–32; Bagrat K. Ioannisiani, "The Soviet 6-meter Altazimuth Reflector Pulkovo Observatory," *Sky and Telescope* 54 (1977):356–62; Henry C. King, *The History of the Telescope* (Cambridge, Mass.: Sky Publishing, 1955); A. B. Meinel, R. R. Shannon, F. L. Whipple, and F. J. Low, "A Large Multiple Mirror Telescope Project," *Optical Engineering* 11 (1972):33–37.

THE LIGHTHOUSE
AND THE SEARCHLIGHT

ANOTHER REMARKABLE ASPECT of the mirror is its ability to project light. If a small intense light is placed at the focus of a parabolic mirror, an approximately parallel beam will be projected to illuminate a distant spot. This made possible the lighthouse, the searchlight, and other means of directed or selective illumination at night.

Twenty-five centuries ago the Chinese made a searchlight by placing a candle or an oil lamp in front of a concave mirror. This was used to light a dark room. Near the close of the fourth century B.C. the *Mo Ching* described the basic features of a light beam projected by a concave mirror.[1] Very little more was written about this phenomenon until 1585 when della Porta alluded to it in his *Natural Magick*. He described a means of seeing at night "what is done afar off." He states that with a concave mirror "we may in a tempestuous night, in the middle of the street, cast the light a great way, even into other men's Chambers. Take the [mirror] in your hand," he instructs, "and set a candle at the point of inversion, for the parallel beams will be reflected to the place desired, and the place will be enlightened above sixty paces, and whatsoever falls between the parallels, will be clearly seen: the reason is, because the beams from the Centre to the circumference, are reflected parallel, . . . and in the place

The Lighthouse and the Searchlight

thus illuminated, letters may be read, and all things done conveniently that require great light."[2]

From della Porta's account, it is clear that the projection of light by a concave mirror had been known and demonstrated for some time. Still, it was not until the seventeenth century that this technique was utilized in a practical manner: as an aid to the seafarer. In ancient times navigation, particularly at night, was often hazardous, and the need for lighting the approach to a dangerous shoreline was self-evident.

The first recorded lighthouse stood on the Pharos of Alexandria. Constructed in the third century B.C., it provided a helpful beacon to ships by burning wood on top of a tower about 450 feet high. Its base was 100 feet square. The lighthouse was located on the then island of Pharos at the west entrance of the Alexandria harbor. Remains of the structure were still visible as late as 1350. In his interesting book on the mirror, Baltrusaitis surveys the history of the recurring legend that the Pharos lighthouse was also equipped with a telescope capable of revealing vessels sixty nautical miles away at sea. There is no evidence, however, that the ancient Egyptians ever constructed lenses or curved mirrors with magnifying eyepieces, both necessary for a telescope.[3]

Protected from the rain or wind simply by a roof, other lighthouses were erected which burned wood. Such general illumination was helpful, but it was useful only for very short distances. Much of the light was wasted in directions from which ships never approached. It needed to be intensified by collecting and projecting it in a desired direction. This is a natural function of the concave mirror.

The Swedes were the first to utilize this in a lighthouse they constructed about 1669, in Landsort, south of Stockholm. Charles XI, king of Sweden, was so fascinated by this new approach that on June 1, 1681, he issued a proclamation granting J. D. Braun a patent for the use of cast steel reflectors for sea lights. In 1687 another lighthouse was built at Örskar in the Gulf of Bothnia. Twelve oil-burning lamps and six concave metal mirrors of 24 inches in diameter were mounted on top of a 148-foot tower.[4] Whether these early projection systems were designed properly is not known, but we may get a clue from a subsequent occurrence.

In 1737 lightning destroyed the Örskar lighthouse. A letter from the king, dated January 2, 1738, ordered it replaced by a stone tower 113 feet high, with a glass, lantern-like enclosure to contain five mirrors of polished steel that were to be "burning mirrors . . . with a parabolic line."[5] This recognition of the correct curve for lighthouse mirrors was unfortunately of no avail in obtaining a maximum intensity of light. Thirty years later it was discovered that of the ten lamps employed, not one had been placed at the focus of each reflector. Instead, a pair of lights was positioned on each side of it. This arrangement could not concentrate the light efficiently and so the benefit of the mirror designer's scientific knowledge was lost.

The first lighthouse in France was built in 1727. It had a curious reflector arrangement. A large cone of wood was covered with shiny tin plates and suspended point-down over a coal fire built on the top of a tower. The design failed because soot deposited on the reflectors and rendered them useless for the projection of light.

The British appear to have been the first to use mirrors effectively in lighthouses. William Hutchinson, a dock-master, was highly motivated in his search for suitable navigational aids as a result of a harrowing experience. The Historical Society of Lancashire and Cheshire records that "in early life Hutchinson was shipwrecked, and the crew being without food drew lots to ascertain who shall be put to death to furnish a revolting and horrible meal to the survivors. The lot fell upon Hutchinson, but they were providentially saved by a ship which hove into sight. He afterwards observed this day as one of strict devotion."[6]

At first, Hutchinson's quest for navigational aids did not include the mirror, for he had little knowledge about it. However, a chance visit to an alehouse in Liverpool revealed its potential to him. It seems he joined a gathering of some "scientific gentlemen of Liverpool," one of whom wagered that he could read a newspaper by the light of a candle thirty feet away. He won by taking a wooden bowl, lining it with putty, and then sticking pieces of looking glass into the putty. This formed a concave reflector which, with the candle optimally positioned in the center, produced a beam of light strong enough to illuminate the newspaper.

Greatly impressed by this demonstration, Hutchinson acquainted himself with the reflecting properties of mirrors, and in 1763 he assembled a successful light-projection system for a lighthouse. He placed a small oil lamp at the focus of a reflector that as nearly as possible conformed to a parabolic curve to obtain the strongest illumination. By 1772 he was instrumental in the construction of four more lighthouses for the channels at Liverpool. Regarding this accomplishment, Hutchinson notes in his book, *A Treatise on Practical Seamanship*, that "we have had and used here in Liverpool reflectors of one two and three foot focus and 3, 5½, 7½, 12 feet diameter. The smallest made of tin plates soldered together, and the largest of wood covered with plates of looking glass, and a copper lamp . . . with the middle of it just in the focus or burning point of the reflector."[7] Though Hutchinson never used his 12-foot-diameter, 3-foot focal-length reflector, which was equipped with a wick 14 inches broad, in any of his lighthouses, he learned that the use of a long focal-length mirror and wide wick was an effective design because it provided good beam spread and kept the reflector away from the smoke of the lamp. The spread of a beam projected from a parabolic mirror may be controlled by the size of the light source and the focal length of the mirror (fig. 32). Hutchinson's work established the first successful reflector and lamp design for sea lane illumination. Soon it was adopted by many other countries.

While night navigation was being improved with the help of the parabolic reflector, the art of flood lighting was introduced by the great French chemist Lavoisier. In 1776 he received a gold medal for his improvement of street lighting in Paris.[8] Rejecting the parabolic mirror because it concentrated the light rays from a lamp in one direction, he recom-

The Mirror and Man

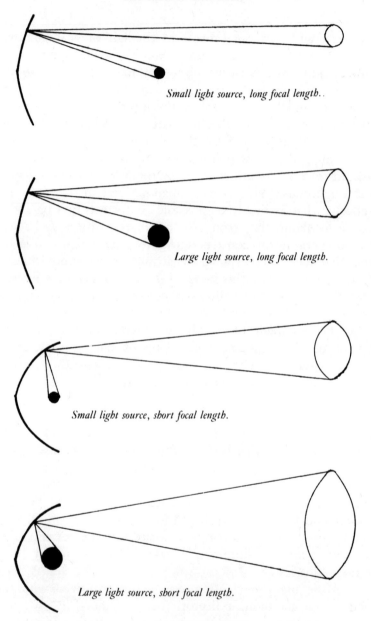

Small light source, long focal length..

Large light source, long focal length.

Small light source, short focal length.

Large light source, short focal length.

Schematics showing increasing beamspreads from four parabolic mirrors as a function of the size of the light source and focal length.

mended the uses of a spherical mirror with a short focal distance. Combined with an oil lamp with a relatively wide wick, this would give a wide and somewhat directed beam of light. His design was implemented by Songrain, a Paris optician, and it replaced the old candle street lights.

The Russians were also at work on light projection systems. In the 1770s a Russian mechanical engineer, I. P. Kulibin, constructed large concave mirrors with light sources. Navigators used these on ships and set them up as beacons. Kulibin constructed his mirrors from small pieces of plane mirrors, gluing them into a clay or cardboard base bent into a parabolic shape. "This mirror," noted the newspaper the *St. Petersburg Vedomosti*, "is particularly advantageous for illuminating large halls and very serviceable for artists and artisans illuminating with bright light a large surrounding area with a minimum expenditure of candles. [A] gallery was lit for 100 yards by this mirror using only 1 candle."[9]

Lighting, particularly in combination with reflectors, was dramatically improved with the introduction in France of Argand's smokeless lamp in 1782. It gave a steady flame and a more intense light. The Argand Lamp consists of a wick, shaped like a hollow cylinder, enclosed in a glass tube. Heat caused by the combustion of oil produces a strong draught, which leads to a greatly increased circulation of air both internally and externally about the wick. This provides more oxygen for a brighter flame. Soon night strollers in Paris and London as well as mariners at sea benefited from this increased lighting. A Frenchman, J. A. Bordier-Marcet, conceived many ideas for utilizing this light with the reflecting properties of the parabolic mirror. About 1802 he rebuilt Argand's workshops in Paris and began construction of superior lamp and reflector combinations that were readily sold for lighting streets and the interiors of buildings. A great advocate of the importance of mirrors in lighting, he spent the next thirty years demonstrating his products at expositions and bombarding municipal and maritime officials and the public at large in France and abroad with pamphlets and

The Mirror and Man

press notices about reflectors. His ingenious mirror and Argand lamp designs were ahead of his time.[10]

Bordier-Marcet was constantly concerned with more efficient light distribution. He noted that when the parabolic reflector with its light source sends out a spreading beam of light from a lighthouse, the upper part of the beam goes skyward and the lower part strikes the water a few hundred yards after it leaves the lighthouse, resulting in much unutilized light. Bordier-Marcet reshaped the parabolic mirror so that the projected beam spread over a greater region horizontally while remaining limited in its vertical spread. It was a combination of two circular mirrors whose horizontal elements only were of a parabolic shape. One was arranged above the light source, the other below it. This was the first light-projection apparatus capable of spreading light equally around the horizon while compressing a large part of the vertical light to contribute more light where it was needed. The first of these star lanterns, as Marcet called them, was tried in 1809. Seamen, much impressed with the increased illumination, called it *"notre salut"* (our salvation). This innovation increased tenfold the power of Argand lamp. It was used very effectively in French harbors, and several of these lanterns were still in service in 1890.[11]

In 1823 Augustin Fresnel introduced a lens system to take the place of the mirror system in lighthouses. Fresnel designed lenses that collected and projected light rays more efficiently than the reflector system, and in a few years they replaced it. Had a powerful light source been available, like that produced after the invention of the electrical generator in 1870, it is unlikely that the mirror light-source combination for lighthouses would ever have been replaced. During the latter part of the century, the Fresnel lenticular system was modified to include both reflectors and lenses to provide even more effective illumination. Further improvement followed when the French first perfected a means of rotating the lamp, lens, and reflector housing 360 degrees to allow a full sweep of the horizon so that the light could be seen by all ships whatever their position in relation to the lighthouse.

The Lighthouse and the Searchlight

The United States did not build a lighthouse with a reflecting system until 1812. Beacons of burning pitch or lamps with candles were used up to that time. The change came about through an invention of Winslow Lewis, an unemployed ship captain.[12] In 1812 the government bought from Lewis, for $20,000, the patent for his "magnifying and reflecting lantern." This consisted of a lamp with a reflector behind and a lens in front. The reflector was made of thin sheets of copper in a spherical form covered with a thin film of silver. Its effectiveness is best described by A. B. Johnson of the United States Lighthouse Service in 1889: "The patentee of 1812 made no pretension to a knowledge of optics . . . and his reflectors came about as near to a true paraboloid as did a barber's basin. Before the lamp was a so-called lens, of bottle green glass . . . 9 inches in diameter which was supposed to have some magnifying power. . . . The effect of the whole was characterized by one of the reporting inspectors as making a bad light worse."[13]

Despite its poor quality, Lewis's reflecting lantern remained in service until about 1840, when new systems were made with heavily silvered parabolic reflectors properly positioned with regard to the light source. By 1852, however, these reflecting systems were replaced by the Fresnel lenticular design so successful in France and generally throughout the world. By 1859 nearly all lighthouses were equipped with Fresnel lenses.

The mirror as a vital part of a light-projection system awaited the development of a more powerful illuminant and efficient ways of producing more accurate reflectors. While the Fresnel projection system was in use, new light sources of greater brilliancy were invented: the limelight and the electric arc. These, combined with a spherical or parabolic mirror, gave birth to the modern searchlight and floodlight. In 1825 Drummond developed the limelight by using an oxyhydrogen flame to bring to incandescence a cylinder of lime that shone with great brilliance. Although it was short-lived because of the advent of the practical electric arc in 1843, its use with a reflector provided light intensities never

The Mirror and Man

before seen after dark. Today's expression "in the limelight" signifies the impact it made. Indeed, the first "searchlamp" used during the Civil War was a Drummond limelight. In the attack on Fort Wagner at Charlestown in the fall of 1863, a Confederate vessel attempting to run the blockade at night was located by a Union warship using a searchlight consisting of a limelight at the focus of a crude spherical mirror.[14]

The electric arc—light produced between carbon electrodes by a current from a voltaic cell—was discovered by Sir Humphrey Davy in 1810. It overtook Drummond's invention in 1848 when the first electric arc lamp for general illumination was installed in Paris.[15] With a mirror, it was used as a searchlight during the Crimean War.[16] This apparently stemmed from the searchlight experiments of Captain Brittes of the French army in 1851. He published his findings in *The Employment of Electric Light in War*. With the invention of the electric dynamo in 1870 as a replacement for the dreadfully cumbersome large batteries and with resultant increase in electric arc brightness because of higher current capability, the mirror was used for the projection of light in most applications in war and peace. Searchlights appeared in the Franco-Prussian War during the siege of Paris in 1870–71, and later they were used in the front lines in the Russo-Turkish War of 1877–78. Beyond military applications, searchlights were used for a variety of other purposes, from the Paris exposition of 1889, where the Eiffel Tower was brilliantly illuminated with searchlights, to ships fitted with projectors making it possible for them to pass through the Suez Canal at night.

In 1886 Johann Sigmund Schukert of Nuremberg produced the first glass parabolic silvered mirror for searchlights.[17] Its greatly increased reflectivity, as compared to the metal mirror, and its more accurate curvature made it a better projector of light. The silvered mirror was adopted for many years as the standard reflector for searchlights and was especially favored by the military, who started with glass searchlight mirrors twenty-four inches in diameter and eventually increased these to sixty inches. The sixty-inch searchlight became the standard for the United States Army in 1909.[18] The intensity (candlepower) of the beam projected by a searchlight with a parabolic reflector depends on the brightness of the light source, the area of the aperture of the mirror, and its efficiency. The efficiency of the mirror depends on how well it reflects and how closely it conforms to the parabolic curve. The 1909 Army sixty-inch searchlight had a peak candlepower of about seventy million candles. Ranges of five thousand yards were attained with these lights.

In England, C. A. Parsons, who founded the Heaton Works, established in 1889 a special department for the production of searchlight mirrors. He introduced many improvements in construction and quality. Adopting Schukert's silvered glass mirror, he protected the silver on the back side with a layer of electroplated copper, further safeguarded with a coat of paint and a backing of sheet lead reinforced with wire netting. It was an extremely important step to provide protection for the silvered surface against salt water, fumes, or other environmental hazards.[19] Like Bordier-Marcet before him, Parsons was a creative mirror designer. One of his searchlight inventions was the paraelliptic mirror, a combination of parabolic and elliptical contours, which was a great improvement over the star lantern. It provides a fan-shaped beam for broad-sector ground illumination while curtailing the vertical light spread skyward and earthward.

A notable advance in searchlight power was made in 1914 when Beck invented the high-intensity carbon arc.[20] This gave the searchlight with its sixty-inch mirror a new military role, that of an antiaircraft weapon eventually providing peak candlepower of 500 million candles with a range of 10,000 yards on a clear night. Its success is illustrated by the record of the 151st Squadron of the Royal Air Force. During August 1918 this group brought down twenty-five German bombers after illumination by searchlight beams, without the loss of a single British plane or aviator.[21]

During the 1930s the silvered glass searchlight mirror was replaced by a metal one. The process of electroforming made possible the fabrication of thick, copper mirror shells of very accurate curvatures coated with the nontarnishing, non-corrosive precious metal rhodium as the reflection surface. Electroforming is an electroplating process permitting the buildup of a thick metal shell upon a mold of a predetermined shape. The surface of the shell in contact with the mold assumes its curvature with great fidelity. For example, a convex parabolic mold is used to build the concave shell of a parabolic reflector. The latter is easily parted from the mold.

During World War II, sixty-inch searchlights were also used for antiaircraft and antisubmarine defense, but increasingly in tandem with the newly invented radar. Radar would search for aircraft at night and, when found, searchlight beams would track the plane for the antiaircraft guns. Although detection of submarines at night on the water's surface by the aircraft radar then available was adequate, some form of illumination was necessary for the final stages of an attack. By equipping aircraft with searchlights housing twenty-four-inch-diameter parabolic mirrors, the RAF was able to reduce substantially the number of submarines leaving their bases nocturnally.[22] Today's military searchlights, with the latest in reflector and light-source design, the Xenon short arc lamps, produce peak intensities up to one billion candles. But the old sixty-inch searchlights are still seen lighting up gala events and shopping malls.

Adaptations of the military searchlights have displaced the Fresnel lens-projection systems in modern lighthouses where high-intensity, narrow beams of light are needed for functions such as channel marking or the illumination of high-hazard areas. One of the first modern lighthouses of this kind was constructed for the Coast Guard at Oak Island, North Carolina, in 1958. A 145-foot-high tower supports a revolving beacon that produces four beams of light, each with a measured peak intensity of seventy million candles. The beacon consists of four thirty-six-inch-diameter parabolic mirrors in searchlight casings mounted on a rotating platform. The Coast Guard has used this design in many of their new lighthouses.

The Lighthouse and the Searchlight

Paralleling the evolution of the searchlight, the flood-light of Lavoisier was used sporadically until the electric arc, and later the filament lamp, were sufficiently developed. It was not until the beginning of the twentieth century that floodlighting as we know it began to be used. In 1907 W. D'A. Ryan installed a large battery of reflectors and arc lamps (floodlamps) at Niagara, illuminating both the American and Canadian Falls. Colored screens in front of the projectors provided pleasing effects that delighted the spectators. Ryan illuminated the tallest building in the country, the Singer Building in New York City, with similar projection systems in 1908. The world's first intensive downtown lighting system, which used these mirror-light combinations, was installed in New Haven in 1911.[23] Since then a myriad of imaginative floodlighting applications have appeared.

NOTES AND SOURCES

1. Joseph Needham, *Science and Civilization in China*, 5 vols. (Cambridge, Cambridge Univ. Press, 1954–74), 4:84.

2. Giambattista della Porta, *Natural Magick* (London: Printed for T. Young and S. Speed, 1658), p. 362.

3. Jurgis Baltrusaitis, *Le Miroir* (Paris: Editions du Seuil, 1968), pp. 160–69.

4. D. A. Stevenson, *The World's Lighthouses before 1820* (New York: Oxford Univ. Press, 1959), pp. 42, 46, 51, 287.

5. Ibid., p. 46.

6. Dudley Witney, *The Lighthouse* (Toronto: McClelland and Stewart, 1975), p. 18.

7. Ibid., p. 19.

8. Stevenson, *Lighthouses before 1820*, p. 190.

9. I. D. Artamonov, *Searchlights and Their Applications* (Moscow: Military Press, Ministry of Defense, 1957), p. 2.

10. Stevenson, *Lighthouses before 1820*, p. 76.

11. Ibid., p. 79.

12. Francis R. Holland, Jr., *America's Lighthouses* (Brattleboro, Ver.: Stephen Greene Press, 1972), p. 14.

13. Arnold B. Johnson, *The Modern Light-House Service of the US Light House Board* (Washington, D.C.: Government Printing Office, 1889), p. 49.

14. F. H. Kohloss, "The Development of Military Searchlights," *The Military Engineer* 22, no. 124 (1930):364.

The Mirror and Man

15. C. A. B. Halvorson, Jr., and R. B. Hussey, "Evolution of Light Projection," *Transactions of the Illuminating Engineering Society* 12, no. 6(1917):249.

16. E. R. Knowles, "Electric Searchlights," *Electrical Engineer* 14, no. 221 (1892):73.

17. Kohloss, "Military Searchlights," p. 365.

18. Ibid.

19. "Glass and Metal Reflectors," *The Engineer* 181, (Feb. 22, 1946):184.

20. Kohloss, "Military Searchlights," p. 365.

21. W. F. Tompkins, "Cooperation of Aeroplane Searchlights in Anti-aircraft Defense," *Military Engineer* 12, no. 63 (1920):266.

22. C. J. Carr, "Aircraft Searchlights for Anti-Submarine Warfare," Symposium on Searchlights, Apr. 15, 1947, *London: The Illuminating Engineering Society*, 1948, pp. 145, 146.

23. Halvorson, "Light Projection," p. 252.

Other sources for this chapter include: William Adams, *Lighthouses and Lightships* (London: T. Nelson, 1871); E. W. Chivers and D. E. H. Jones, "The Function and Design of Army Searchlights," Symposium on Searchlights, Apr. 15, 1947, *London: The Illuminating Engineering Society*, 1948, pp. 7–25; Benjamin Goldberg and Nels C. Benson, "Analysis of Para-elliptic Reflector," *Journal of the Optical Society of America* 39 (1949):497–500; Henry E. Haferkorn, "Searchlights: A Short Annotated Bibliography of Their Design and Their Use in Peace and War," *Professional Memoirs, Corps of Engineers, U.S. Army,* Jan.-Feb. 1916, pp. 118–28, and Mar.-Apr. pp. 250–63; H. S. Hele-Shaw and D. MacKenzie, "A New Metallic Mirror for Searchlights," *Scientific American Supplement*, July 25, 1908, p. 55; W. Helmore et. al., "High Power Searchlights Capable of Wide Divergence," Symposium on Searchlights, Apr. 15, 1947, *London: The Illuminating Engineering Society*, 1948, pp. 26–49; Benjamin W. Norregaard, *The Great Siege: The Investment and Fall of Port Arthur* (London: Methuen, 1906); J. E. Wesler, "The Coast Guard Lights a New Lighthouse," *Illuminating Engineering Society* 54 (1959):-415–416.

THE MIRROR
IN SPACE

O N JULY 20, 1969, at 9:56:15 P.M., Eastern Standard
Time, Neil Armstrong, commander of Apollo 11,
stepped on the moon. He and Edwin Aldrin planted several
devices on the moon's surface, including a solar wind detec-
tor, a seismometer, and a mirror array consisting of a panel
of retroreflectors. Millions watched on television as Aldrin
removed the special mirror unit from the bay of the Eagle
and carried it about sixty feet away from the craft. Then
Armstrong adjusted the array so that its face was approximately perpendicular to an imaginary line from the landing
site to the earth. The first man on the moon oriented a pallet
structure of 100 high-precision corner-cube retroreflectors assembled in an eighteen-inch-square array (fig. 33). Regardless of the orientation of a retroreflector, a light beam incident on it will be reflected back on itself. Thus, if you held
your eye directly behind a flashlight shining into such a
mirror from any angle, the return beam will always enter
your eye. Manufactured of fused silica to minimize deformation by the large temperature changes on the moon, each
corner cube had its sides perpendicular to each other to
within the phenomenal accuracy of less than 0.3 seconds of
arc. The front surface of each corner cube (fig. 34) was
ground to a 1½-inch-diameter circle, and each of the sides

Panel of retroreflectors on the moon. (NASA photo)

The Mirror in Space

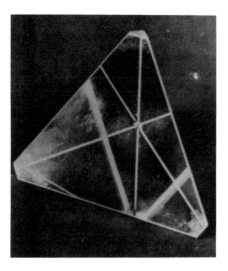

Basic retroreflector.

was polished to within about two-millionths of an inch of perfect flatness.

These mirrors were put on the moon as part of the Laser Ranging Retroreflector Experiment (LRRE). Its purpose was to measure as precisely as possible the distance from the earth to the moon. Since 1957 conventional radar techniques had been used to determine this distance to an accuracy of 0.7 mile, but this is not precise enough for scientific purposes. Now a laser pulse generator coincident with a large astronomical telescope, for example, the 107-inch-diameter mirror of the McDonald Observatory, is aimed toward the vicinity of the bank of reflectors on the moon. The emitted laser pulse is reflected from the retroreflectors directly back to the telescope and the transmit time measured. Because speed of light is known, the distance between the earth and moon can be determined.

It is interesting that a cylindrical mirror forms an integral part of the laser system to greatly increase the efficiency of the laser output. It concentrates the light from an electronic flash lamp onto a lasing material, like a ruby rod, that, in turn, emits a laser beam.

This technique showed so much promise that a second LRRE was deployed by the Apollo 14 mission in February

1971, and a third by the Apollo 15 mission in July 1971. The mirror array set down by the last mission was a much larger one. It consisted of a hinged two-panel assembly, one of 204 reflectors, and the other of 96 reflectors. Range information provided from three points on the moon attained accuracies within plus or minus three inches over the greater than 200,000-mile distance. Daily measurements made from the McDonald Observatory and other parts of the world have been able to plot minute changes in the earth-to-moon distance which allow studies of gravitation, relativity, and earth and lunar physics.

By determining the minute variations in the orbit of the Moon, the LRRE will reveal the gravitational interaction of the earth, moon, and sun. This will allow a greater understanding of the nature of gravity and will help to answer the question of whether the force of gravity is slowly diminishing. Using the moon as a reference point, scientists can study the strange wobbling of the earth on its axis known as Chandler's Wobble.

Another major objective of the LRRE is to learn more about the drifting of the continents on the earth. Recent theories indicate that these move with respect to each other at the rate of up to four inches per year. The lunar distance measurement will give the longitudes of the observing stations with such accuracy that the expected motions could be observed in two or three years after stations are established in Hawaii and Australia in 1984. Accurate information about the lunar orbit will be made available. The wobbling motion of the moon about its center of gravity may be accurately measured, and this will provide a greater understanding of the internal structure of the moon and may reveal whether the moon is a fragment of the solar system or whether it was formed in outer space and then caught in the earth's gravity. The National Aeronautics and Space Administration (NASA) scientists and their academic colleagues believe that definitive answers to these and other questions will be available by 1988.

In July 1976 NASA launched an earth-orbiting, mirror-

The Mirror in Space

clad satellite known as the Laser Geodynamic Satellite (LA-GEOS) in a 3,600-mile-high orbit. The surface of this 900-pound, spherical, 24-inch-diameter satellite is covered by 426 retroreflectors (fig. 35). These corner-cube reflectors are of the same high accuracy as those used in the LRRE system, some of whose work it is duplicating as a check. Laser beams from a worldwide network of ground stations are bounced off these mirrors, and the resulting range measurements can detect continental drift as small as two centimeters. A better understanding of earthquakes may also be obtained by measuring the rate and the amount of movement across selected faults, such as the San Andreas in California.

These examples of mirrors in space are mere hints at

Laser geodynamic satellite. (NASA photo)

their possible extraterrestrial use as envisaged by Hermann Oberth, the pioneer thinker who as early as 1923 imagined, among other things, a giant space mirror. This mirror would provide directed sunlight on selected parts of the earth to carry out different kinds of tasks. Commenting in 1957 on his early ideas, which were at first viewed with benign humor, he stated: "I am certain that my space mirror will one day be a reality. The critics object to its size; they cannot conceive that such a structure 100 kilometers (60 miles) in diameter with an area of 70,000 square kilometers (27,000 square miles) could be built in space. They forget that in space, and in all that concerns space travel, dimensions are entirely different from those known on the Earth."[1] The mirror he describes consists of hexagonal modules, each several miles across, the whole being in continuous rotation. The mirror consists of a metal frame over which thin, highly reflective metal foil, about 0.002 inches thick, is stretched. Because no appreciable forces are acting on it, it can be made large and light.

Oberth believed that it would take ten years or more to build this immense mirror. If the raw materials could be obtained from an asteroid instead of bringing them up from earth, progress would be faster. He realized that this was a gigantic task, but he believed that it would be no more difficult than any other space project. The pressure of sunlight as it strikes and reflects from it would rotate or manipulate the position of the mirror in space. It is true that the pressure of light from an electric bulb is infinitesimal but the pressure of sunlight on a large surface like Lake Erie, which is 9,940 square miles for example, amounts to several tons. This idea also lies behind the Solar Sail project.

Because Oberth's space mirror design consists of movable hexagonal modules, he speculated that light energy reflected from the sun could be concentrated at specific locales on the earth or spread over wide areas. With this mechanism, he visualized large towns being brilliantly lighted at night with individual facets of the mirror. Icebergs dangerous to shipping could be melted with directed concentrations of sunlight. Icebound sea routes to Spitzbergen and northern Siberia could be kept open, and the climate of the arctic regions could be improved. In the temperate latitudes, rapid drops in temperature and the sudden return of cold weather in the spring could be prevented. Measures could be taken to counteract night frosts in the autumn, which would be a boon for fruit and vegetable crops in many countries. Oberth was aware of the huge cost of building a 100-kilometer-diameter mirror in space. He knew it might cost billions of dollars, but he argued that that was less than the cost of a small war and but a fraction of the cost of the world wars which have drained modern man's resources.

After the first satellite success by the Russians in 1957, Oberth's ideas became more realistic. The first mirrors to appear in space were part of orbiting telescopes. This was a natural extension of earthbound telescopes, whose ability to see into space is limited by the turbulence and absorption of the atmosphere. Free of the atmosphere, a telescope can see at least ten times farther into space. And it can see X-rays, gamma rays, and ultraviolet rays that are absorbed by the atmosphere. On December 7, 1968, NASA launched the Orbiting Astronomical Observatory, designated AOA-2, to measure ultraviolet light from stars and nebulae. This satellite carried nine telescopes ranging from eight to sixteen inches in diameter. The AOA-2 transmitted data on thousands of stars during its first year in operation.

Two years later the UHURU, another satellite observatory, was launched to gather new data on existing X-ray stars and to look for new ones that were found in the Milky Way and other galaxies. This was followed by the AOA-3, which contained a telescope with a thirty-two-inch-diameter mirror, and the 1973 Skylab, which used six telescopes to study the sun. The latter has provided over 30,000 pictures of the solar disk. It has also recorded profound changes in the structure of the corona (the hottest and outermost region of the solar atmosphere) during flares. Further studies of the sun were carried out with reflector telescopes on the Orbiting Solar Observatory, designated OSO-7. The instruments of OSO-7 viewed the turbulent corona continuously in the intense light of the sun, which can be observed from the ground only during brief moments of total eclipses because of the interfering bright sunlight that is scattered by dust and gas in our atmosphere. Solar flares are of special interest because of their great energy, their effects on earth, and the remarkable plasma phenomena that occur in them, for example, the nuclear reactions that occur in the large flares.

Under construction in 1984 is the Space Telescope, ST. It will be launched from the Space Shuttle in 1986. Featuring a ninety-five-inch-diameter reflector, it will be the first major observatory in space that can be repaired and refurbished by the shuttle while in orbit about 350 miles above the earth. With the ST it should be possible to investigate the structure and history of the cosmos and to determine whether the universe was born in a giant explosion as proposed by the big bang theory or in some other manner.

Another space project that uses the mirror is the solar sail, an idea also proposed by Oberth. In 1976 NASA revived his idea when it began to think about economical ways to power a craft on journeys through the solar system. The following year NASA awarded design contracts for material and booms for a square, 160-acre sun sail as a step toward the construction of immense solar sailing ships in space. One design (fig. 36) uses a mirrored plastic surface about 0.0001 inch thick to reflect the sun's rays and ultralightweight extensible booms for the spars and masts of the solar sail. The furled sail would be taken up by the Space Shuttle, unfurled on the shuttle platform, and erected in space by astronauts. The huge, mirrorlike sail would reflect sunlight, the light photons exerting a windlike pressure to propel a 2,000-pound spacecraft containing a scientific payload. Like the earthbound sailboat, the system would not require fuel. By tacking against or with the solar photon stream, the solar sailcraft could fly in any direction. In 1984 the solar sail project was being held in abeyance because of funding and development problems, but it remains a viable approach to a transportation vehicle in space.

The Mirror in Space

A solar sail. (NASA photo)

Recently, scientists have presented some brilliant ideas for the use of mirrors in space as a solution to the earth's energy crisis. After all, for mankind at large, the most important aspect of space is the almost uninterrupted supply of solar energy that might be harnessed by such mirrors. In a paper before the American Institute of Aeronautics and Astronautics and the American Astronomical Society in 1975, G. R. Woodcock and D. L. Gregory proposed capturing solar energy with a satellite covered with twenty-two square miles of mirror surface. They believe that this immense mirror system could provide 10,000 megawatts of useful power, about twice the present hydroelectric capability of the Grand Coulie Dam, by the conversion of solar energy to electricity within the satellite. This could then be transmitted to power stations on earth in the form of directed microwaves.[2]

Another provocative space mirror idea was presented in a study by the Congressional Subcommittee on Space Science and Applications considering future space programs in 1975. Conceived by K. A. Ehrike, it deals with space light for night illumination.[3]

Ehrike points out that mankind requires more light than ever before. We consume more and more energy for that purpose with each passing year. Now he believes that "space technology has put Man at the threshold of establishing his own light reflectors in space and [sic] to illuminate at least a small area in relation to the globe, but a fairly large area by customary standards with a brightness that may vary from the equivalent of one full Moon . . . to the intense brightness of the Sun."[4] He proposes a three-tier program of increased illumination on selected areas on the earth by orbiting mirrors in the sky. The first program, called Lunetta, would utilize an orbiting reflector to direct sunlight to earth to provide a low light level for night illumination by the year 2000. The second program, called Soletta, would create about one-half the sun's brightness and would provide sufficient light during the night to stimulate food growth by enhancing photosynthesis production. This might be ready by 2035. The largest of them all, the third program, is Total Soletta, which would provide the same brightness as the sun. This might be finished by 2040.

Lunetta would consist of a modularized orbiting solar reflector, rather like a huge spoked wagon wheel with reflective sheeting between the spokes. There would be an area of five square kilometers made up of a cluster of mirrors 500 meters in radius. Orbiting the earth geosynchronously at about 8,000 miles, it could light an area of 88,000 square kilometers to an illumination of twenty full moons, which would be sufficient for intense urban and traffic lighting. The angle of the reflector would be adjustable so as to maintain the light on a localized area throughout the night. One-quarter of the mirror area would be sufficient for either crime control or public safety in urban areas or for sufficient community lighting in developing countries. Each square kilometer of Lunetta would provide illumination equivalent to that produced by the consumption of thirteen million barrels of oil annually. Of special note is the fact that where no area lighting system exists, such as in developing countries, it eliminates the need for electrical distribution systems and the associated consumption of metals.

The Soletta system, which would provide illumination equal to about 40 percent of the sun's intensity, would be much larger. For example, a reflector area of 1,080 square kilometers in an orbit of 4,000 miles above the earth would provide an illuminated area of 2,800 square kilometers with about four hours of illumination. This orbiting system would most likely consist of a large cluster of wheel-shaped modularized reflectors in a linear array tilted to merge as a single beam. It could increase agricultural production by interrupting the night with a period of illumination to stimulate photosynthesis, and it could provide an extension of daylight for agricultural areas in regions with short summers. These mirrors in space could provide an increase in food production of $220 million (in 1975 dollars) for each 100 square kilometer of Soletta area.

A Total Soletta would depend on a proportionate increase in the linear array of orbiting solar reflectors. This would provide selected areas on earth with an extra sun whose energy could be used for large-scale food production, large-scale fresh-water generation or desert irrigation, and direct power generation for industries and cities.

As daring as these ideas appear, they are surpassed by those presented in "Orbiting Mirrors for Terrestrial Energy Supply," a proposal delivered at the third NASA Conference on Radiation Energy Conversion, January 1978, by K. W. Billman, W. P. Gilbreath, and S. W. Bowen.[5] This goes beyond Oberth's dreams. It is a proposal for a system of many small, lightweight, orbiting reflectors that would provide continuous exposure of the sun's rays to a worldwide system of terrestrial energy conversion sites. This system is called SOLARES, the Solar Energy System.

Terrestrial solar energy programs that use mirrors for electric power generation are limited by the small inherent solar energy density (0.25 kilowatts per square meter), daily variation in the sun's intensity, and clouds and their changing position in relation to the sun. These systems are under-utilized. Frequently, they can operate for no more than six hours a day without, or twelve hours a day with, storage.

The SOLARES, or orbiting mirrors notion, seeks to solve these problems by providing continuous and slightly concentrated reflected solar energy to selected terrestrial conversion sites. This additional energy would be about one kilowatt per square meter as compared to the best average obtained in the United States, which is one-fourth that amount. For a given solar farm output of electrical power, its acreage could be reduced by five times, thus eliminating the need for mirror collectors, land photovoltaic converters, maintenance, and the like. Costly energy storage can also nearly be eliminated. The authors suggest, with supporting data, that their system can sizably augment and possibly even replace the present electrical generating capacity of the world.

Billman, Gilbreath, and Bowen propose a multitude of relatively low-orbit mirrors (fig. 37), each with computer for inclination and orbit control. These mirrors project their so-

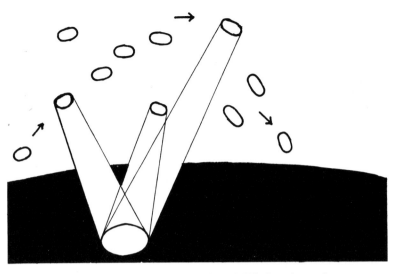

Schematic of the SOLARES focusing mirror array.

lar images on a common ground station beneath them. There
would be a number of these stations around the world. Each
disk-shaped mirror with a radius ranging from one-half to
two kilometers would consist of a plastic film about 0.0001
inch thick with an evaporated aluminum reflecting surface.
These mirrors would be rotating slowly as they orbit around
the earth so that the reflected image of the sun can remain
fixed on the site. The mirrors would orbit at an altitude of
approximately four thousand kilometers circling the earth
every three hours. The mirrors would be built on an earth-
orbiting space station at an altitude of about one thousand
kilometers and then solar-sailed to the desired orbit. This
procedure would enable operators to modify orbits as
needed. A mirror might be lowered for repair and maintenance or raised to increase the power output of SOLARES if
required.

In order to supply an amount of electrical energy in
excess of that currently used worldwide, 8,000 mirrors, each
of 1.6 kilometer radius, would have to be distributed. These
would provide continuous one kilowatt per square meter
average radiant energy to five selected ground sites. If the
entire radiant input were converted to electrical energy with
a conversion efficiency of 15 percent, one site could produce

The Mirror and Man

an average of about 200,000 megawatts of power compared with the current United States usage of 250,000 megawatts. Five such sites located around the world would exceed current world needs, which are estimated to be about one million megawatts.

The presence of so many mirrors in space traveling in so many orbits may seem overimaginative but, "no physical constraints, such as the availability of orbital space or the ability to track and control such a number, have surfaced which would preclude this number."[6] Unlike a few equivalent larger mirrors, groups of these smaller ones would be easier to manufacture, cost less, and be simpler to deploy, control, and maintain.

SOLARES could allow us to conserve nonrenewable fossil fuels now used to generate electrical power. Additional savings might come from its use for water desalination, in industry, crop drying, and space heating. SOLARES would also help remove pollution due to mining, transporting, and burning of fossil fuel and danger of pollution from nuclear energy. SOLARES would require a $500 billion investment, or an annual investment of $33 billion for fifteen years. This is not excessive when one considers that from $25 to $30 billion are needed each year to expand a conventional power system to meet the demands of the United States alone. The proponents of SOLARES conclude that "it should be appreciated that the investment for the orbiting mirror system by the U.S. . . . would ultimately allow our nation to regain its position of an energy exporter to the World rather than ever-increasing dependency on foreign import."[7]

These applications of the mirror in space may enhance life on earth, but there remains the great dream of man, the colonization of space. Here the mirror can contribute by providing controlled sunlight for man's personal, agricultural, and industrial needs in the sealed confines of a space colony.

Gerard K. O'Neill of Princeton University proposed such a colony in 1973. He considered several designs, all of them surrounded with a great mirror, and selected finally what is now called the Stanford Torus (fig. 38). Here, colo-

Stanford Torus. (NASA photo)

nists would live in a ring-shaped tube that is connected by six large access routes, or spokes, to a central hub where incoming spacecraft dock. The spokes provide a way into the living and agricultural areas, which are located in the tube. This structure would orbit the earth in the same orbit as the moon in a stable position equidistant from the earth and the moon. To simulate the normal gravity of the earth, the huge, wheellike structure would rotate at one revolution per minute about the central hub.

Directly above this hub there is a curved lightweight mirror over half a mile wide. It floats above the colony. Small rockets keep its position properly adjusted. It is angled toward the sun to reflect light down onto a double ring of secondary mirrors. These, mounted in rings around the hub, are supported by the spokes, and they reflect the light into the structure through a set of louvered mirrors. These mirrors can also be tilted to turn the sunlight away. This is done twice a day, at dusk and dawn. With the help of this reflected sunshine, the colonists might be able to raise enough food for themselves on only 156 acres. The abundant solar energy would foster agriculture of unusual productivity and generate electricity to power solar furnaces in the hub, which could refine aluminum, titanium, and silicon from lunar ore shipped through space. These materials would, in turn, be used to fabricate new structures that could probe deeper into the universe.[8]

NOTES AND SOURCES

1. Hermann Oberth, *Man into Space*, tr. G. P. H. DeFreville (New York: Harper, 1957), pp. 97, 98.

2. G. R. Woodcock and D. L. Gregory, "Derivation of a Total Satellite Energy System," AAIA paper 76-640, AIAA/AAS Solar Energy for Earth Conference, Los Angeles, Apr. 24, 1975.

3. Krafft A. Ehrike, "Space Industrial Productivity: New Options for the Future" (Space Light Program), Presentation to the Committee on Science and Technology and the Subcommittee on Space and Applications, Hearings on Future Space Programs, July

22–30, 1975, Washington, D.C., Govt. Printing Office, pp. 150–83.

4. Ibid., p. 150.

5. K. W. Billman, W. P. Gilbreath, and S. W. Bowen, "Orbiting Mirrors for Terrestrial Energy Supply," from *Radiation Energy Conversion in Space*, vol. 61, *Progress in Astronautics and Aeronautics*, Prepared by the 3d NASA Conference on Radiation Energy Conversion at NASA Ames Research Center, Moffett Field, Cal., Jan. 26–28 1978, Published by the American Institute of Aeronautics and Astronautics, 1978, pp. 61–80.

6. Ibid., p. 69.

7. Ibid., p. 76.

8. Gerard K. O'Neill, "The Colonization of Space," *Physics Today*, Sept. 1974, pp. 32–34.

Other sources for this chapter include: Ivan Bekey, "Space and Energy," in *A New Era in Space Transportation, Proceedings of the 27th International Astronautical Congress*, Anaheim, Calif., Oct. 10–16, 1976 (New York: Pergamon Press, 1977); Frank S. Child, "Laser Reflectors on the Moon," *Glass Industry* 50 (1969):429–33; James E. Faller and Joseph E. Wampler, "The Lunar Laser Reflector," *Scientific American*, Mar. 1970, pp. 38–50; T. A. Heppenheimer, *Colonies in Space* (Harrisburg, Pa.: Stackpole Books, 1977); Michel Hoez and Marc Foex, "Remarks concerning Solar Furnaces in Space," *Solar Energy* 13 (1972):417–20; Keith McBee, ed., "Earth's Crustal Movements to be Studied," *NASA Activities* 10 (Nov. 1979):10–11; Mission Evaluation Team of NASA Manned Spacecraft Center Apollo 11 Mission Report, *Laser Ranging Retroreflector Experiment—Section 11: The Lunar Surface* (Washington, D.C.: NASA-SP-238, 1971); NASA Technical Paper 2147, *The NASA Geodynamics Program: An Overview* (Washington, D.C.: Office of Space Science and Applications, NASA, 1983); Press Kit Release no. 76–67, "Project Lageos," *NASA News;* Jesco von Puttkamer, "Humanization beyond Earth: The New Age of Space Industrialization," Presented at the AAAS Annual Meeting, Feb. 12–17, 1978, Washington, D.C., Symposium on "Prospects for Life in the Universe: Ultimate Limits to Growth."

THE
SUBJECTIVE MIRROR

Despite its importance in the scientific pursuits of the nineteenth and twentieth centuries, the mirror continued to retain its grip on man's psyche. To the creative minds of this period, the mirror, symbolically or figuratively, has continued to convey in different ways a feeling of eeriness. In a more general sense, it has served as the screen of man's projections of his identity, his uncertainties, and his desires. Psychologists have begun to study this and educators to benefit by it. The mirror continues to provide a way toward self-examination.

Many writers of romantic literature in the nineteenth century were fascinated by the idea of "reflection as the soul," or "double," a topos of antiquity. E. T. A. Hoffmann, a German writer of great imagination, is considered the classical creator of the double projection. In his *Story of the Lost Reflection* (1814), the hero sells his mirror reflection (his immortal soul) to a devil for the love of an evil seductress. Soon after he receives the mockery and contempt of the world as a man without a reflection. Finally, he is forced to spend the rest of his life searching for his soul.

In subsequent literature dealing with one's loss of soul through a similar action, the double is usually a symbol of the character's past, his evil tendencies and his death rather than future immortality. Sometimes the double appears as an outpouring of self-condemnation. The hero, conscious of his guilt, transfers the responsibility for certain personal actions to his other self, the double. His inordinate fear of death leads to a transference to the double. In *Monsieur du Miroir* Nathaniel Hawthorne wrote of his mirror double with a touch of narcissism, and he felt uneasy about his reflection's power to haunt him after his death because of his vanity. In a turnabout idea, Bram Stoker's vampire, Dracula, is a soulless creature who has no reflection in a mirror.

The double motif has been important in Russian literature. At the turn of the twentieth century, the Russian symbolists used this theme. Probably the most prominent was Andrei Bely (1880–1934), who wrote numerous poems and several works where the reflected image plays an important role. As with other writers, in Bely's case the reflection portended a tragic end for his characters. In his most important works, such as *The Second Symphony, The Third Symphony*, and *In Petersburg*, the narratives are full of references to mirrors and mirror images. This followed in the tradition of earlier Russian writers who used the mirror as a device to describe the subjective worlds of their characters. For example, Gogol's *The Nose* and Dostoyevski's *The Double* use the mirror in this way.

The double phenomenon, a subject of superstition and fiction, became, early in the twentieth century, the focus of a comprehensive study by Otto Rank. Rank, a renowned Austrian psychoanalyst, published *The Double* in 1924. His basic premise is that man's need for self-perpetuation, for immortalizing himself, resulted in the concept of the double, or soul, which in turn led to the development of civilization and its spiritual values.

Of interest was the motivation for his study, a motion picture about a mirror double. In the early days of the cinema, Rank saw the movie *The Student of Prague*, a visual drama by Heinz Ewers that was not unlike E. T. A. Hoffmann's *Story of the Lost Reflection*. The hero, Balduin, gives up his reflection to the evil one, Scapinelli, who in turn provides the former with wealth and affluence to assure his marriage to Margit, a young innocent daughter of a royal family. During the course of the movie, Balduin's reflection appears in many independent and ungraceful attitudes and actively interferes in the hero's romance with Margit. Balduin tries to escape his insidious reflection but fails. In despair, Balduin, on the point of committing suicide, prepares his loaded pistol and starts to write his last will and testament. But, again, his double stands grinning before him. In a frenzy Balduin fires at the phantom, which disappears at once. He laughs with relief and, believing that he is now rid of his tormentor, uncovers his hand mirror which had been wrapped in cloth and views himself for the first time in a long while. In that moment he feels a sharp pain in his chest, sees blood on his shirt, and realizes that he has been shot. In the next instant he collapses and dies. The smirking Scapinelli appears, in order to tear up the contract over the corpse.

This visible portrayal of psychological events crystallized for Otto Rank, through the deception of the motion picture, what the written word could not do and permitted him to focus on the real meaning of an ancient theme that had become obscure or misunderstood in its course through history. Rank observed that the double represented the problems of man's relationship to himself. This also was the preoccupation of the Russian symbolist writers in their mirror stories. It was connected with an urge to turn back their mode of thinking in order to reclaim their souls, which nineteenth-century science had seemed to destroy. The mirror became a symbol of man's rejection of materialistic reason in his search to repossess his lost soul.

Many writers have created different mirror worlds of their fantasies, fears, or frustrations. In his mirror world of *Through the Looking-Glass*, Lewis Carroll reflects his kind of topsy-turvy. He replaces the normal left-right reversals of a mirror image to front-back inversions, time reversals, and sense-nonsense reversals. Alice has to walk backwards to approach the Red Queen. Later, the White Queen explains the advantages of living backward in time and the Unicorn remarks, "You don't know how to manage Looking-glass cakes. Hand it round first, and cut it afterwards." The king himself, reflects this sense-nonsense inversion when he says that he wished he had eyes good enough to see Nobody.

Jean Cocteau believed that poetry is to be found in everything, and so he shunned the prosaic. Viewing prose as an inversion of poetry, his vivid imagination likened the former to the "mirror world," an inversion of the "real world." He feared the mirror world of prose and portrayed this concern allegorically in his stage play (and film) *Orphée*. The poet, Orpheus, in search of his wife, passes through a mirror into the Kingdom of Prose, which for him is unbearable compared to his worldly poetical existence. He sees this mirror world as an expressionless void, a world of death cruelly indifferent in its dreary continuity. Early in the play Cocteau's concern is enunciated by the guardian angel Heurtibise, who tells Orpheus as they approach the mirror, "I am letting you in to the secret of all secrets, mirrors are gates through which death comes and goes. Moreover, if you observe your whole life in a mirror you will see death at work as you see bees behind the glass in a hive."[1]

Cocteau's friend Jean Genet was born in squalor of unknown parents. A homosexual and a thief, he has spent many years in prison. Life to him was a struggle for self-identity and the frustration in not finding himself. The mirror symbolized this for him. Genet expressed his feelings about mirrors in *The Thief's Journal* when he describes how, at the funfair in Antwerp, he once watched Stilitano, his cruel mentor and lover, trapped in the hall of mirrors (a mirror maze). A group of bystanders who could see him roared with laughter as Stilitano tried desperately to get out, bouncing off the mirrored glass looking for an exit. His curses, which could not be heard, were seen as wild grimaces. He was trapped in the mirrored labyrinth, a lonely figure fighting a frustrating battle.

To Genet this represented the frustration of the human situation. He saw his friend cut off from all contact and communication by invisible walls of glass, and his only direct experience of reality was that of bumping against his own reflection. It objectified to Genet his own view of man's eternal conflict with the shadows of his own identity. Genet portrayed this frustration in a number of his works, including the ballet *Adame Miroir* and his play *Le Balcon*. According to Richard N. Coe, Genet may have conjectured that if appearance is the basis of reality, then the image in the mirror is, in the end, more real than the object that causes it. And, being more real, it attracts to itself all those characteristics of the living: independent actions, a past and a future identity. To Genet,

> the mirror image which symbolizes the Self-that-knows is superior to the real Self that merely exists; the reflection in the glass is not simply an irrefutable fact of existence, but the object of an intense emotional relationship, which is at the same time love and hate; love because the Image is, when all is said and done, oneself, the essential element without which no self-knowledge would be conceivable . . . hate, because the double is indispensable to the Self, and, at the same time in its quality of conscious Otherness, superior.[2]

Many experiments have been undertaken by psychologists to understand the self-image. One of the most interesting of these was carried out by L. H. Schwarz and S. P. Fjeld and reported in 1968.[3] Possibly drawing from the scryer's handbook, they carried out their experiment with their subjects late in the evening in a dimly lit room to heighten the productivity of illusionary effects. They used sixty-four subjects of which sixteen were normal, sixteen neurotic, sixteen sociopathic, and sixteen psychotic. Each group had eight men and eight women. The subjects, in turn, were seated facing a sixteen-inch-square mirror placed two feet away, with the only illumination provided by a small tape recorder bulb located three feet behind them. The experimenter was seated by the tape recorder. After allowing the subject to get used to his surroundings, he was asked to concentrate on his image in the mirror and to report freely on what he saw, felt, or thought during a thirty-minute session. All his remarks were recorded.

Most subjects experienced gross distortions of their perceived reflections in a multitude of strange ways. Many also associated their experiences during their tests with significant persons or important events in their lives. A normal patient remarked, "That image reminds me of Molière. . . . He used to dislike physicians." A terrifying illusion was described by a sociopathic male: "I am thinking of the opera of Leoncavallo, *Il Pagliacci.* I'm seeing the figure of the main actor with a bleeding heart in his hand. . . . The audience is laughing . . . but he does not know that his wife is unfaithful to him. . . . now an isolated finger from an amputated hand . . . a piece of dirty finger nail . . . reminds me of a girl friend who had an abscess in her finger."[4]

In general, the abnormal groups experienced the most unusual perceptual distortions including experiences of fear, unusual physical symptoms, projected feelings of aggression, and erotic fantasy. While normal groups occasionally experienced similar imagery, their perceived reflections were often of a tranquil and sympathetic nature. The experimenters emphasized that the high rate of fantasy formation was due to features of hypnotic induction present in the experiment. Such procedures do facilitate fantasy production and remind us of the mirror-gazing experiences of the scryers. They certainly verify Max Dessoir's statement that, in considering the power of the mirror to image strange things and events, "the most important factor [is] the person that [sees] and not the instruments of seeing."[5]

The mirror often projects an image modified by the way we believe we appear. This not only includes the bodily surface, contour, and posture of the observer but also his attitudes and emotions. The body image seems remarkably plastic. Attempts were made by psychologists to measure this amorphous perception of one's self, that is, the difference between a person's perceived image and his true image. It appeared impossible to have a subject peer into a mirror and describe his bodily image in an objective manner. He

The Mirror and Man

could only report what he saw, which was influenced by his subjective attitudes. This difficulty was largely overcome with a special kind of mirror. In 1964 Traub and Orbach described a new instrument, the Adjustable Body Distorting (ABD) Mirror, created to explore a person's mental perception of his physical appearance.[6] A special flexible full-length mirror was built which could be manipulated so that one's image could be varied from a condition of extreme distortion to one of no distortion. The mirror, of excellent optical quality, about fifty inches high by twenty inches wide, could be bowed either into a concave or convex shape. The degree of bowing was accomplished by controlled pressure applied at the edges. The amount of distortion in the mirror was registered on four dials individually geared to the four distorting motors providing the pressure along each edge. With the mirror distorted, a subject was positioned before the mirror and asked to vary its shape until he believed he looked like himself.

The procedure employed is informative. The subject is confronted with a grossly distorted reflection of himself. He stands erect, seven feet from the mirror and is given four successive trials with different initial distortion configurations. On the first trial, he appears tall, with a pin head, large elongated body and legs tapering to tiny feet. On the second, he appears short with an enormous head and tapering legs. On the third, the subject is angularly distorted with one side of his body projecting laterally. The fourth view involves a short body with dwarfish legs. The subject is told, "Your task is to adjust the reflection of yourself so that it looks just like you." He is shown how to manipulate the two switches that change the mirror shape and is given all the practice he needs to master the task. Moreover, he is offered as much time as is necessary to produce an accurate reflection. His last instruction is: "If you make the proper adjustments, you will end with an accurate reflection of yourself. However, if you should find that, no matter what you do, the reflection never looks exactly like you, just stop when the reflection looks most like you." When the adjustments have been completed and the subject stops, a record is made of the final adjustment and the subject is asked, "Are you satisfied that the reflection you see in the mirror looks exactly like you? . . . Are there any parts of your reflection that don't quite resemble you?"[7] All replies are recorded.

Traub and Orbach collected data on a sample of thirty persons, none of them mentally deficient. Occasionally, the subjects showed exaggerated reactions to viewing themselves in the mirror. Some were unable to look at the mirror at all. Others suffered dizziness, nausea, headache, and obsessive preoccupation with detail. One surprising result, however, was common among many of the subjects. After spending some minutes adjusting the mirror, they would sheepishly declare that they had forgotten precisely what they looked like. Some urgently requested that they be allowed to examine themselves in a normal mirror before proceeding with the test, which they were allowed to do.

The results of these tests disproved a basic assumption: that a normal person would choose one reflection that best represented the mental picture of his body. Instead, a rather wide range of reflections seemed to be equally acceptable. The surprised investigators reexamined these unexpected results. Invariably, when each of the subjects was shown a variety of views of himself, all within his range of acceptability, he readily acknowledged that he could discriminate one from the other but stated that they were all perfectly good reflections of his body. This was contrasted with the response of the subjects when shown a series of photographs taken within their acceptability range. Here, the subjects responded by accurately identifying which photographs were distorted and selecting the one undistorted photograph from the series. It appeared that each subject needed an external reference point, the undistorted reflection, to be able to identify accurately the distorted reflections. Without this, they often stated they did not remember exactly what they looked like. The investigators found that judgment of the head and shoulders was the most accurate and least variable. Next in consistency were judgments involving assymetric distortions of the vertical halves of the body. The least accurate were those judgments involving legs and feet. This seems logical since we learn how our body looks largely from mirrors and photographs, both of which frequently show only the head and shoulders.

In 1972 R. F. Wilps published a doctoral dissertation which expanded on the Traub and Orbach experiments.[8] He argues that the mirror cannot provide objective test results. Because of its power to call forth profound and emotionally laden attitudes about the self, it arouses a complex pattern of behavior in the subject's reaction to his reflected image. The mirror might better be used like the ink blot test as a projective technique where the subject describes the images he sees and the associated ideas that come to mind. Wilps suggested that various bodily and verbal sensations of subjects be recorded while they carry out particular activities before the ABD mirror. Obese people, those with physical deformities, or those with psychological disorders, could in this way provided behavior pattern models of great value. An analysis of a patient's mirror behavior patterns may provide valuable clinical information about his actual, as opposed to his ideal, self-image, and his ability or lack of it to literally face himself.

Wilps concludes his paper with an intriguing thought: "To be able to control and distort [one's reflection] is an opportunity to recreate oneself in one's own fancied image; an opportunity which has fascinated all of the subjects of the mirror study. To measure that distortion and record a person's responses to it, may well be the closest we can yet come to mapping the soul."[9]

Daniel Cappon in his book *Eating, Living, and Dying,* noted that very fat and very thin persons have distorted self images. They may look at themselves fleetingly or in some preferential manner and do not appear to themselves as they objectively appear to others. But if the fat or thin person stands and looks at his reflection studiously in a full-length mirror on a regular basis, he will rapidly correct his distorted image.

In his successful practice, Dr. Cappon requires his pa-

The Subjective Mirror

tients to view themselves for one minute, twice a day. Employing the three sided, angled mirrors used in fitting rooms of clothing stores, the patients take a full measure of their contours, study the rolls of fat or the angular ridges of bones where rounded curves should be. Very soon, the reflection they see in the mirror becomes the self image they carry about for a full day. They associate their self image with their distorted approach to food and act to free themselves from this.[10]

Contrary to a superstition that the very young should be prevented from looking into a mirror to avoid a narcissistic fixation, such an action can be most beneficial. This was first noted by the psychologist Fritz Wittels, who wrote,

> When I was still a small boy, I woke one day with the overwhelming realization that I was an "I", that I looked externally, to be sure like other children but nonetheless was fundamentally different and tremendously more important. I stood before the mirror, observed myself attentively and often repeatedly addressed my image by my first name. In doing so, I evidently intended to create a bridge from the image in the external world over to me, across which I might penetrate into my unfathomable self. I do not know if I kissed my reflection, but I have seen other children kissing theirs; they come to terms with their ego by loving it.[11]

Here the mirror has served the healthy function of identifying the child to himself.

Today the mirror is being used in schools as a positive courage builder for the very young. It is of particular value in the kindergarten class where many children first arrive frightened and insecure. The teacher must reassure the child by instilling a feeling of physical and mental well-being and an attitude of "you can do it." These traits must be installed as soon as possible since the schoolchild's educational patterns are crystallized by the second or third grade. The teacher exploits the fact that most of us can rationalize our appearance in a mirror in a most favorable manner, finding specific aspects in our reflections that make us attractive, handsome, debonair, or rugged.

The importance of this technique has been recognized for a long time by the affluent. Robert Coles in his *Privileged Ones*, a study of American children of well-to-do and rich parents, notes that these children are exposed to many mirrors in their homes. "They have mirrors in their rooms, large mirrors in adjoining bathrooms. When they were three or four they were taught to use them" for inspection of their personal appearance to insure neatness and cleanliness as a means of nurturing self-esteem. Coles, who had spent many earlier years studying underprivileged children, remarks that "with none of the other American children I have worked with have I heard such a continuous and strong emphasis put on the 'self.' "[12] This healthy ego characteristic fostered by the positive mirror image is reflected in the assurance and authority the "privileged ones" exhibit as leaders in society.

When man became a self-conscious creature, the mirror so stimulated his imagination about his being that he endowed it with supernatural qualities. It became an object of enchantment that provided an answer to his basic concerns. As a result, many cultures adopted the mirror as a reverential artifact or symbol. It became an earthly representation of a supreme deity or a bridge between man and his God. Man's conception of his universe, bounded by his religious beliefs, changed when the mirror provided mankind with an objective view of reality. In so doing, the mirror became disenchanted. It freed man of his earlier misapprehensions about himself. It has also contributed to his understanding of the universe. The modern mirror, looking outward, contributes to the broadening of scientific and technological horizons. But it remains, as well, a powerful tool of introspection, a guiding metaphor for distinguishing between outward appearance and inner truth.

The Subjective Mirror

NOTES AND SOURCES

1. Elizabeth Sprigge and Jean-Jacques Kehm, *Jean Cocteau: The Man and the Mirror* (New York: Coward-McCann, 1968), p. 186.
2. Richard N. Coe, *The Vision of Jean Genet* (New York: Grove Press, 1968), p. 18.
3. L. H. Schwarz and S. P. Fjeld, "Illusions Induced by the Self-Reflected Image," *Journal of Neurological and Mental Diseases* 146 (1968):277–84.
4. Ibid., p. 281.
5. Max Dessoir, "The Magic Mirror," *The Monist* 1 (1890):87.
6. A. C. Traub and J. Orbach, "Psychophysical Studies of Body Image," *AMA Archives of General Psychiatry* 11 (1964):53–66.
7. Ibid., p. 61.
8. Ralph F. Wilps, Jr., "Body Image, Perceptual Accuracy, and the Adjustable Body-Distorting Mirror," Ph.d. diss., Adelphi University, Garden City, N.Y., 1972.
9. Ibid., p. 72.
10. Daniel Cappon, *Eating, Living, and Dying: A Psychology of Appetites* (Toronto: Univ. of Toronto Press, 1973), p. 103.
11. Otto Rank, *The Double*, tr. and ed. Harry Tucker, Jr. (Chapel Hill: Univ. of North Carolina Press, 1971), p. 81, n. 22.
12. Robert Coles, *Privileged Ones* (Boston: Little, Brown, 1977), p. 380.

Other sources for this chapter include: Seymour Fisher, *Body Consciousness* (Englewood Cliffs, N.J.: Prentice-Hall, 1973); Martin Gardner, *The Annotated Alice* (New York: C. N. Potter, 1960); William C. Jordan, "Mirror Image," *Grade Teacher*, Jan. 1971, p. 100; Oleg A. Maslinikov, "Russians Symbolists: The Mirror Theme and Allied Motifs," *Russian Review* 16 (1957):42–52; J. Orbach, Arthur C. Traub, and Ronald Olson, "Psychological Studies of Body Image," *AMA Archives of General Psychiatry* 14 (1966):41–47; Robert Rogers, *The Double in Literature* (Detroit, Mich.: Wayne State Univ. Press, 1970); Paul Schilder, *The Image and Appearance of the Human body* (New York: International University Press, 1950).

Index